A Student Guide to Energy

A STUDENT GUIDE TO ENERGY

Volume 1: Oil, Natural Gas, Coal, and Nuclear

John F. Mongillo

AN IMPRINT OF ABC-CLIO, LLC
Santa Barbara, California • Denver, Colorado • Oxford, England

Copyright 2011 by John F. Mongillo

All rights reserved. No part of this publication may be reproduced, stored in a retrieval system, or transmitted, in any form or by any means, electronic, mechanical, photocopying, recording, or otherwise, except for the inclusion of brief quotations in a review, without prior permission in writing from the publisher.

Library of Congress Cataloging-in-Publication Data

Mongillo, John F.
  A student guide to energy / John F. Mongillo.
    v. cm.
  Includes bibliographical references and index.
    Contents: v. 1. Energy : oil, natural gas, coal, and nuclear — v. 2. Solar energy and hydrogen fuel cells — v. 3. Wind energy, oceanic energy, and hydropower — v. 4. Geothermal and biomass energy — v. 5. Energy efficiency, conservation, and sustainability.
    ISBN 978-0-313-37720-4 (set hard copy : alk. paper) —
ISBN 978-0-313-37721-1 (set ebook) — ISBN 978-0-313-37722-8
(v. 1 hard copy: alk. paper) — ISBN 978-0-313-37723-5 (v. 1 ebook) —
ISBN 978-0-313-37724-2 (v. 2 hard copy : alk. paper) —
ISBN 978-0-313-37725-9 (v. 2 ebook) — ISBN 978-0-313-37726-6
(v. 3 hard copy : alk. paper) — ISBN 978-0-313-37727-3 (v. 3 ebook) —
ISBN 978-0-313-37728-0 (v. 4 hard copy : alk. paper) —
ISBN 978-0-313-37729-7 (v. 4 ebook) — ISBN 978-0-313-37730-3
(v. 5 hard copy : alk. paper) — ISBN 978-0-313-37731-0
(v. 5 ebook)   1. Power resources — Encyclopedias. I. Title.
  TJ163.16.M66   2011
  621.042—dc22   2011000481

ISBN: 978-0-313-37720-4
EISBN: 978-0-313-37721-1

15 14 13 12 11   1 2 3 4 5

This book is also available on the World Wide Web as an eBook.
Visit www.abc-clio.com for details.

Greenwood
An Imprint of ABC-CLIO, LLC

ABC-CLIO, LLC
130 Cremona Drive, P.O. Box 1911
Santa Barbara, California 93116-1911

This book is printed on acid-free paper ∞

Manufactured in the United States of America

CONTENTS

Acknowledgments	*ix*
Introduction	*xiii*

Chapter 1   Energy and Energy Sources   1
     Energy's Role in Our World   1
     Energy Policies Need to Be Addressed   2
     History of Energy in the United States   3
     What Is Energy?   5
     Energy Can Be Converted from One Form to Another   6
     Today's Sources of Energy   8
     Electricity, a Secondary Source of Energy   16
     How Is Energy Measured?   18
     Worldwide Consumption of Energy   18
     Worldwide Uses of Energy   19
     Future World Energy Consumption   21
     Interview
         Linda Currie   28

Chapter 2   Petroleum   39
     How Do We Use Petroleum Today?   39
     Petroleum, a Dominant Energy Source   40

	History of Petroleum	40
	What Is Petroleum?	41
	How Did Petroleum Form?	41
	Searching for Oil	42
	Drilling for Oil	43
	Crude Oil Refineries	46
	The Main Products of Petroleum	47
	U.S. Production of Crude Oil	54
	Crude Oil Imports to the United States	54
	OPEC	56
	Oil Shales and Oil Sands	57
	U.S. Imports Other Than Refined Crude Oil	59
	The Strategic Petroleum Reserve	60
	Environmental Issues	60
	How Much Oil Is Left?	62
	The Future of Petroleum	63
	Interviews	
	Jason Diodati	49
	Keats Moeller	64
Chapter 3	**Natural Gas**	**71**
	World Consumption of Natural Gas	72
	History of Natural Gas	73
	How Is Natural Gas Formed?	74
	The Contents of Natural Gas	74
	Locating Natural Gas Deposits	76
	Drilling for Natural Gas	76
	Delivery of Natural Gas	77
	How Is Natural Gas Measured?	78
	Global Natural Gas Reserves	79
	World Gas Production Countries	81
	The World's Largest Consumers of Natural Gas	82
	Liquefied Petroleum Gas—Propane	85
	Natural Gas Vehicles	85
	Natural Gas Benefits	93
	Natural Gas Emissions	93
	Future of Natural Gas	94
	Interview	
	Bob Walters	86

Chapter 4 Coal — 103
- Coal: A Major Source of the World's Energy — 103
- History of Coal — 104
- Major Uses of Coal — 104
- What Is Coal and How Is It Formed? — 106
- Where Is Coal Found? — 106
- How Is Coal Mined? — 106
- How Is Coal Transported? — 112
- How Does a Coal-Fired Power Station Produce Electricity? — 113
- Major Coal-Producing States — 114
- Major Global Coal-Producing Countries — 115
- Exporters of Coal — 116
- Coal Importers — 117
- Coal Consumption — 118
- Environmental Issues — 119
- Clean Coal Technology — 121
- The Future of Coal — 128

Chapter 5 Nuclear Energy — 131
- Why the Interest in Nuclear Energy? — 132
- What Is Nuclear Energy? — 133
- History of Using Nuclear Energy — 134
- Concerns about Nuclear Weapons — 135
- International Atomic Energy Agency — 135
- World Use of Nuclear Energy — 136
- Nuclear Fuel — 138
- The Kinds of Nuclear Reactors — 142
- Some Major Nuclear Reactor Manufacturers — 145
- Nuclear Power Plants Generate Waste — 145
- Cutting-Edge Nuclear Technologies — 151

Interview
- Dr. Charles Ferguson — 154

Books and Other Reading Materials — *159*

Government and Nongovernmental Organization Web Sites — *165*

Energy Data — *169*

Energy Time Line: 3000 B.C. To A.D. 2009 — *197*

Profiles — *203*

Opportunities in Renewable and Nonrenewable Energy Careers	209
Energy Product Developers and Manufacturers	213
National Science Education Standards, Content Standards	217
Index	219
About the Author	263

ACKNOWLEDGMENTS

First and foremost I would like to thank David Paige, formerly Acquisitions Editor, Health and Science, of ABC-CLIO/Greenwood for his support and effort in molding the energy series into its current form. Thanks to the staff at Apex CoVantage for the project management, copyediting, and proofing services, and Ellen Rasmussen, Senior Media Editor, for her photo research contributions.

Much of this series would not be possible without the efforts of the Green Advocates who provided interviews describing their go-green activities and their enthusiasm for this series. The Green Advocates included Ross McCurdy, High School Science Teacher, Ponaganset, Rhode Island; Linda Currie, Energy Outreach Coordinator, Bay Localize, Oakland, California; Jason Diodati, Chemistry Teacher, Marc and Eva Stern Math and Science School, East Los Angeles, California; Bob Walters, Technology Education Teacher, DeWitt Middle School, Ithaca, New York; Mark Westlake, High School Physics Teacher, Saint Thomas Academy, Mendota Heights, Minnesota; Bhavna Rawal, High School Science Teacher, Northbrook High School, Houston, Texas; Rande Gray, Design Project Manager, Hannaford Supermarkets, Maine; Rick Peck, Science Teacher, Seneca Ridge Middle School, Sterling, Virginia; Stephanie Harman, Science Teacher, Maumee Valley Country Day School, Toledo, Ohio; Tom Traeger, Science Teacher, La Cañada High School, La Cañada, California;

Mary E. Spruill, Executive Director of the National Energy Education Development (NEED); Michael Arquin, Director, KidWind Project, St. Paul, Minnesota; John W. Lund, Director of the Geo-Heat Center at Oregon Institute of Technology, Klamath Falls, Oregon; Phillip Cantor, North-Grand High School, Illinois; Sister Susan Frazer, RSM, MSW, St. John Bosco Boys' Home, Jamaica, West Indies; Don Carmichael, Science Teacher, Adlai E. Stevenson High School, Illinois; Mr. and Mrs. Gerald McGrath, Massachusetts; and Marie Norman, Principal, Westwood Elementary School, Zimmerman, Minnesota (The Westwood Elementary School was the first LEED-certified school in Minnesota).

The publisher and the author are pleased to have received permission to reprint interviews with members of the Spirit Lake Wind Project and the National Energy Education Development (NEED). The NEED Careers in Energy interviews included David Chen, Program Manager for the TXU Energy Solar Academy for TXU Energy, Dallas, Texas; Dr. Charles Ferguson, Philip D. Reed Senior Fellow for Science and Technology, Council on Foreign Relations, Washington D.C.; and Keats Moeller, Senior Advisor of Recruiting and Staffing, ConocoPhillips Company, Houston, Texas.

The author wishes to acknowledge and express the contribution of the many government and nongovernmental organizations and corporations who provided assistance to the author in the research for this energy series.

A special thanks to the following organizations that contributed technical expertise and resources, photos, maps, and data: Government organizations and their representatives included the Department of Energy, Office of Energy Information Administration, Office of Fossil Energy, Environmental Protection Agency, Bureau of Reclamation, National Renewable Laboratory Agency (NREL), the NREL's National Wind Technology Center, National Oceanic and Atmospheric Administration (NOAA), and the National Aeronautic Space Administration (NASA). Thanks to other organizations, including the Alliance to Save Energy, Sandia National Laboratories, American Wind Energy Association (AWEA), Geothermal Energy Association, National Gas Association, Nuclear Energy Agency, American Coal Council, and the National Association for Stock Car Auto Racing (NASCAR).

In addition, the author wishes to thank Amy Mongillo, George F. McBride, and Paula Jutkiewicz for their proofreading and typing support and Edward and Rachel Patrick and Sister María Elena Cervantes,

RSM, for their friendship and support, too. Special accolades to Carolyn Koeniger, Peter Mongillo, and Jane and Gareth Phillips, who provided invaluable resources such as video Web sites, bibliographies, government and nongovernment Web sites, science activities, energy timelines, and much more.

In conclusion, please note the responsibility of the accuracy of the terms is solely that of the author. If errors are noticed, please address them to the author so corrections can be made in future revisions.

INTRODUCTION

We cannot simply think of our survival; each new generation is responsible to ensure the survival of the seventh generation. The prophecy given to us, tells us that what we do today will affect the seventh generation and because of this we must bear in mind our responsibility to them today and always.

—*Great Law of Peace of the Haudenosaunee
(Six Nations Iroquois Confederacy)*

Presently, energy drives the global economy by producing much of the goods and services manufactured and sold in the marketplace. The global supply and demand for energy influences the major stock markets in all of the capitals of the world. Energy impacts all of our lives by supplying the means for transportation, electricity, manufactured goods, and agricultural production. Therefore, any disruption in the energy supply system or shortages of energy resources will have a major impact on the economies of all the countries in the world.

The present energy system provides us with many benefits, but it also impacts and degrades our environment. Fossil fuel supplies will also be running out before the end of the 21st century. Therefore, a global sustainable energy program that includes renewable energy sources, energy conservation policies, and energy efficiency programs is needed.

World governments, nongovernmental organizations, international companies, universities and colleges, entrepreneurs, and citizens are debating present and future energy policies. A few of the questions they are addressing include the following:

- How will fossil fuel shortages, the depletion of nonrenewable energy sources, and the rising costs of fuels, such as petroleum, impact the world's population of energy consumers, particularly those living in developing countries?
- Will all proposed energy policies and programs be sustainable in format to balance the future energy needs and demands of people without damaging the environment?
- How reliable, efficient, and affordable are the renewable energy sources, such as wind, solar, and hydrogen, that are to power the future and replace traditional energy sources?
- What implementation plans are best to conserve energy in homes, businesses, transportation systems, and agricultural production?

The consumption of fossil fuels continues to increase the world greenhouse gas emissions and global temperatures. One estimate is that 76 percent of global warming is caused by carbon dioxide alone. As atmospheric temperatures rise, global temperatures also rise, causing global warming. These atmospheric conditions cause the potential for major climate change that may not be reversible.

There is no question that topics concerning energy resources and technology will continue to be in the news and play a major role in economics, public policy, science, ethics, and political and environmental issues in the 21st century.

THE *STUDENT GUIDE TO ENERGY* SET

A Student Guide to Energy is a multivolume reference set and an excellent research tool for developing a working knowledge of basic energy concepts and topics. The set provides an interdisciplinary perspective on the study of energy. Coverage of traditional nonrenewable energy and conventional sources includes petroleum, natural gas, coal, and nuclear fission. The renewable, or alternative, energy sources covered include solar energy, wind power, geothermal power, hydropower, tidal power, biomass and biofuels, and hydrogen fuel cells.

No one book can keep track of all the changing events and developments in the energy field or even hope to present the most current information about each issue. There is too much going on in the energy research field to document all events or issues in one set. However, *A Student Guide to Energy* provides an excellent tool for developing a working knowledge of energy-related topics that are important to understanding our present and future needs for energy resources and energy efficiency.

Organization

A Student Guide to Energy is divided into five volumes.

Volume 1: Oil, Natural Gas, Coal, and Nuclear. Volume 1 highlights our present dependence on the nonrenewable energy sources such as petroleum, natural gas, and coal that provide the majority of the world's energy needs. The last chapter reports on nuclear energy. Interviews, suggested video sites, science activities, and a bibliography complement each chapter in the volume.

Volume 2: Solar Energy and Hydrogen Fuel Cells. In volume 2, solar energy and hydrogen fuel cells are presented as alternative, renewable energy sources. There are many U.S. schools using solar energy. The hydrogen economy is discussed in chapters 4 and 5. Interviews, suggested video sites, science activities, and a bibliography complement each chapter.

Volume 3: Wind Energy, Oceanic Energy, and Hydropower. Wind energy, hydropower, and tidal energy are presented in volume 3. Interviews, suggested video sites, science activities, and a bibliography complement each chapter.

Volume 4: Geothermal and Biomass Energy. Volume 4 reports on geothermal energy and geothermal heat pumps. Chapters 4 and 5 report on biofuels and biomass as energy resources. Interviews, suggested video sites, science activities, and a bibliography complement each chapter.

Volume 5: Energy Efficiency, Conservation, and Sustainability. The last volume in the set, volume 5 focuses on the importance of living in sustainable society where generation after generation does not deplete the natural resources or produce excessive pollutants. Energy conservation, energy efficiency, and energy sustainability are covered.

Additional topics, including carbon and ecological footprints and global warming issues, are also covered. Interviews, suggested video sites, science activities, and a bibliography complement each chapter.

Special Features of the Five-Volume Set

- **Biographies.** Men and women who have made contributions in the energy field and in energy technologies.
- **Interviews.** Firsthand reporting of teachers, professors, and business owners who play a prominent role in the go-green energy field.
- **Career information.** Suggested careers to assist young people to explore the possibilities of a go-green career in energy-related fields.
- **Energy companies and organizations.** A listing of web sites of the major corporations that are involved in cutting-edge research and in the development of energy technology for the future.
- **University and college resources.** Energy resource links and web sites from schools and colleges.
- **Government and nongovernmental resources.** Web sites for all of the major government agencies and nongovernmental agencies that are conducting energy research and funding.
- **Science activities.** Suggested student research activities at the end of each chapter in the volume.
- **Video sources.** More than 100 approved video web sites intermeshed within the text for the introduction and enrichment of the chapter content that is covered.
- **Energy time line of events.** Important energy and energy technology milestones.
- **Bibliography.** Book titles and articles relating to the subject area of each chapter, presented at the end of each chapter for additional research opportunities.
- **School energy news.** Several go-green U.S. schools have installed and use renewable energy resources. These resources include photovoltaics, geothermal energy and geothermal heat pumps, and wind power. The teacher interviews discuss how energy projects are part of their science and math studies. These projects include building biodiesel autos and pickup trucks and even a 100 percent electric-powered car.

- **National Science Education Standards.** The content in *A Student Guide to Energy* is closely aligned with the National Science Education Standards. *A Student Guide to Energy* does not fall into a single traditional discipline but rather supports learning in a range of disciplines, including physics, chemistry, biology, mathematics, engineering, and technology.
- **Hundreds of illustrations.** Diagrams, photos, charts, and tables that enhance the text and provide additional information for the reader.

A BRIEF OVERVIEW OF PRESENT AND FUTURE ENERGY RESOURCES

Nonrenewable Energy Sources

Petroleum

Presently, 90 percent of the world's energy is derived from the consumption of coal, petroleum, and natural gas. According to government reports, fossil fuels will continue to be the major source of energy for the transportation, industrial, and residential sectors. For example, the world's demand for petroleum will have increased to 91 million barrels per day by 2015, from 85 million barrels per day in 2006. By 2030, consumption will have reached 107 million barrels per day. Overall, global energy consumption is projected to grow by 44 percent over the 2006 to 2030 period.

Ten countries produced 60 percent of total world production of oil. Following are the top five, which produced 42 percent of the world total, and their share of total world production:

- Russia, 13 percent
- Saudi Arabia, 12 percent
- United States, 7 percent
- Iran, 5.4 percent
- China, 5.1 percent

Following are the top five exporting countries, accounting for 59 percent of U.S. crude oil imports in 2009:

Canada, 1.854 million barrels per day
Mexico, 1.177 million barrels per day
Saudi Arabia, 1.021 million barrels per day

Venezuela, 0.803 million barrels per day
Nigeria, 0.673 million barrels per day

Natural Gas

According to government studies, worldwide natural gas consumption will increase to 158 cubic feet in 2030, from about 100 trillion cubic feet in 2005. Natural gas will probably replace petroleum and coal wherever possible. The reason is that natural gas combustion produces less carbon dioxide than coal or petroleum production and products. Therefore, natural gas is expected to remain a key energy source for the industrial sector. Today, natural gas is used extensively in residential homes, commercial buildings, and industrial plants in the United States. In fact, it is the dominant energy used for home heating. Natural gas supplies nearly one-fourth (23%) of all of the energy used in the United States, with more than 66 million homes in the United States using it. The use of natural gas is also rapidly increasing in electric power generation and cooling.

Worldwide, natural gas remains a key energy source for the industrial sector and for electricity generation. The biggest consumers of natural gas in 2005 were the United States, Russia, Germany, and the United Kingdom. However, since 2000, the demand for natural gas in Spain had grown by 92 percent, putting Spain in sixth place in Europe, behind the United Kingdom, Germany, Italy, France, and the Netherlands.

Coal

Coal accounts for approximately 49 percent of electricity output in the United States. It is the world's most abundant and widely distributed fossil fuel. Although coal deposits are widely dispersed, more than 59 percent of the world's recoverable reserves are located in five countries: Australia, China, India, United States, and Canada. The world's largest producers and consumers of coal are China, Poland, Russia, India, and the United States. Major hard-coal producers include China, the United States, India, Australia, South Africa, Russia, Indonesia, Poland, Ukraine, and Kazakhstan.

According to a study by International Energy Outlook, coal's share of world energy consumption is projected to increase by 29 percent by 2030. Coal's share of the electric power sector will reach 46 percent in 2030. China is the world's largest coal producer, accounting for nearly 28 percent of the world's annual production and about 70 percent of China's total energy consumption.

Nuclear Energy

In 2010, President Barack Obama announced an $8.3 billion federal loan to build two new reactors in Georgia. "We'll have to build a new generation of safe, clean nuclear power plants in America," said President Obama. The United States is still the largest single producer of nuclear energy in the world, with 104 units supplying more than 750 billion kilowatt-hours. This is a 25 percent increase in total power over the course of 15 years, as a result of improving equipment, procedures, and general efficiency, without a new reactor order. (As of 2010, Watts Bar Unit 1, finished in 1996, was the latest completed U.S. reactor.)

According to the Nuclear Energy Agency, as of 2009, France had the second-largest number of commercial reactors with 59, and it was building one new reactor at Flamanville, with plans for another new reactor at Penly. France is a major global producer of nuclear power for electricity. France's first nuclear reactor began operating in 1974, and the most recent reactor prior to Flamanville came into use in 2000. About 78 percent of France's electricity is produced by nuclear energy. France is a major exporter of electricity to other countries in Europe.

Renewable Energy Resources

Solar Energy

Presently, several solar technologies have been developed to use the sun's energy as renewable energy resource for heat and electricity. The major technologies include photovoltaic cells, concentrating solar power systems, and special solar collectors for space heating and hot water.

Photovoltaic (PV) cells, made of semiconductors such as crystalline silicon or various thin-film materials, convert sunlight directly into electricity. According to Vicki Mastaitis of the Interstate Renewable Energy Coalition, more than 400 schools in the US now have PV systems on their buildings. The typical grid-tied PV system installed in a school is one or two kilowatts.

In fall 2009, President Barack Obama visited the DeSoto Next Generation Solar Energy Center in DeSoto County, Florida. The solar plant, located in the southwest area of Florida, has more than 90,500 photovoltaic cells that can generate 40,000 megawatts of electricity. Other states are also exploring solar power, including Michigan, California, Texas, Utah, New York, and Colorado

In all, more than 80 countries are making plans to use solar energy as part of their renewable energy portfolio, which also includes wind power, biofuels, geothermal energy, tidal power, and wave power. As of 2010, China is the world's leading manufacturer of solar cells; it claims to have more than 400 PV companies and manufactures approximately 18 percent of the photovoltaic products worldwide. Additionally, there are now more than 300,000 buildings with PV systems in Germany. Spain is a major country investing and installing solar energy as well, and Brazil, Italy, Korea, India, Taiwan, and Saudi Arabia are developing solar energy projects.

Concentrating solar power (CSP) technologies use special-shaped mirrors to reflect and concentrate sunlight onto receivers. The solar energy is converted to heat in the receiver. This heat energy then is used to produce steam that powers a steam turbine or heat engine to generate electricity. The Department of Energy states that CSP could be a major contributor to solving our nation's energy problems now and in the future.

According to the National Renewable Energy Laboratory, Acciona Energy's Nevada Solar One is the third-largest CSP plant in the world and the first plant built in the United States since 1999.

Overseas, in 2009, Spain installed the largest solar tower in the world. The 500-foot-high solar tower, located near Seville, Spain, has the capacity to supply electricity to 10,000 homes.

Solar water heaters are another innovation. The state government of California has approved a $350 million program to subsidize the installation of solar water heaters to help reduce greenhouse gas emissions. Today, many countries use solar hot-water systems for a wide variety of purposes, including for household needs and for heating swimming pools.

Solar hot-water heating systems are very popular in countries with plenty of daylight solar radiation. Some of these countries include Cyprus, Israel, Greece, Japan, Austria, and China, the latter of which is the number one user of solar water heaters. At least 30 million Chinese households now have solar hot-water heaters. In 2009, the country accounted for approximately 80 percent of the world's market for solar hot-water heaters.

According to the Department of Energy, solar water heaters, also called solar domestic hot-water systems, can be a cost-effective way to generate hot water for your home. They can be used in any climate, and the fuel they use—sunshine—is free.

Today, many experts believe that a major switch to solar energy is the best answer to reducing fossil fuel use and emissions. Many solar energy companies in the United States and around the world are researching, planning, and using technologies to harness the sun's energy to generate electricity for businesses, homes, schools, and large communities.

Fuel Cells

The United States and other countries are continuing to explore fuel cell technology and applications because of its benefits. "The fuel cell industry in 2007 reported that there had been substantial job growth and gains in sales and research," according to the Worldwide Industry Survey. Fuel cells are clean, efficient, and economical.

A fuel cell is a device that uses hydrogen (or hydrogen-rich fuel) and oxygen or other fuel to create electricity through an electrochemical process. According to the Department of Energy, there are several types of fuel cells currently under development, each with its own advantages, limitations, and potential applications. They include polymer electrolyte membrane (PEM) fuel cells, direct methanol fuel cells, alkaline fuel cells, and phosphoric acid fuel cells.

Presently, hydrogen fuel cells are used in a variety of ways. Fuel cells are now powering bicycles, boats, trains, planes, scooters, forklifts, and even buses. Police stations, hospitals, banks, wastewater treatment plants, and telecommunication companies use fuel cells for cellular phones and radios.

The world's leading automakers are working on alternative technologies using fuel cells for cars, buses, and trucks. According to Allied Business Intelligence, "The current $40 million stationary fuel cell market will grow to more than $10 billion by 2010. Fuel cells are currently being developed in sizes appropriate for use in homes and other residential applications."

Wind Power

In 2008, the United States became one of the fastest-growing wind-power marketplaces in the world. That year, wind power accounted for approximately 40 percent of all new U.S. electricity-generating capacity. The Department of Energy reported that wind power could generate 20 percent of all U.S. electricity needs by 2030.

The global picture for countries using more wind power looks very promising. The Worldwatch Institute estimates that wind energy could easily provide 20 to 30 percent of the electricity needed by many countries, and the development of wind power technology is not unique to the United States. Many countries are developing this renewable energy resource. As an example, according to the American Wind Energy Association (AWEA), Denmark leads the world, producing more than 20 percent of its electricity needs at home from wind energy.

Most economists predict that the largest growth markets for wind turbines are in Germany, India, Spain, Great Britain, and China. In 2010, China became the number one manufacturer of wind turbines. But let's look at Europe: Europe is high on wind power. In fact, wind turbines generate more electricity in Europe as an alternative source of energy than in the United States. In the early twenty-first century 40 percent of the world's wind farms will be in Europe. In addition to wind farms, Europeans are encouraged to invest in wind-power installations for their homes and businesses in an effort to conserve energy resources.

Hydropower

In Norway, hydroelectric power meets more than 90 percent of the country's electricity needs. Presently, hydroelectric power plants produce about 24 percent of the world's electricity. This is enough electricity to supply more than 1 billion people with electrical power for their household needs. "The world's hydroelectric power plants, output a combined total of 675,000 megawatts, the energy equivalent of 3.6 billion barrels of oil," according to the National Renewable Energy Laboratory.

Much of the electricity used in Brazil, Canada, Norway, Switzerland, and Venezuela is generated from hydroelectric power plants. These countries generate more than 170,000 megawatts of electricity. That is an enormous amount of energy—enough power to support the electrical needs of more than 110 million households in the United States.

Some of the major hydroelectric power dams in the world, include the Three Gorges Dam in China, the Itaipu Dam on the border of Paraguay and Brazil, and the Guri Dam in Venezuela.

Tidal Power Energy

Many countries are examining the potential to harness tidal energy to drive turbines for electricity. However, only a few sites in the world have been identified as possible tidal power stations. Presently, tidal power stations are operating in France, Canada, Russia, and China. The largest is the one in France.

Although much of the electricity produced in France is from nuclear power plants, the country has a tidal power plant as well. The Rance tidal power plant is in operation on the estuary of the Rance River, in the northwest corner of France. The power plant went online in 1966 and became the world's first electrical generating station powered by tidal energy. The plant produces 240 megawatts of power. Canada, China, and Northern Ireland are developing tidal energy plants as well. Presently, Nova Scotia's tidal power plant uses the Bay of Fundy tides to produce enough electricity for 6,000 nearby homes.

Geothermal

Presently, geothermal energy is the fourth-largest source of renewable energy in the United States, where about 3,000 megawatts of geothermal electricity are connected to the grid. According to the Department of Energy, energy generated from geothermal sources accounted for 4 percent of renewable energy–based electricity consumption in the United States. The United States continues to produce more geothermal electricity than any other country, making up approximately 30 percent of the world's total. And two countries alone, the United States and the Philippines, together account for 50 percent of the world's use of geothermal energy. As of August 2008, geothermal capacity in the United States totaled nearly 3,000 megawatts, produced in several states such as Alaska, California, Hawaii, Idaho, Nevada, New Mexico, and Utah. California alone produces more megawatts of geothermal energy than any country in the world.

Biomass and Biofuels

In 2009, as part of the ongoing effort to increase the use of domestic renewable fuels, U.S. Secretary of Energy Steven Chu announced plans to provide $786.5 million from the American Recovery and Reinvestment Act to accelerate advanced biofuels research and development and to provide additional funding for commercial-scale biorefinery demonstration projects.

Global biofuel production tripled between 2000 and 2007 but still accounts for less than 3 percent of the global transportation fuel supply. However, global demands for biofuels are expected to more than double between 2009 and 2015, according to a new global analysis released.

Major new contributors to the growth of global biofuels between 2009 and 2015 will include Indonesia, France, China, India, Thailand, Colombia, Malaysia, Philippines, and Argentina.

Energy's Future

Most energy experts believe that at least midway through the twenty-first century we will continue to depend heavily on fossil fuels for transportation and electricity needs. Therefore, it is necessary to be more efficient in using these energy sources.

However, energy conservation and energy efficiency are not enough to cut the growth of emissions. To get deeper reductions, more clean and renewable energy sources must be used.

As we look into the future, we need to inspire our young people, who hopefully will be more involved in being energy-efficient, exploring hands-on green energy projects, and investigating and shadowing careers in go-green vocations.

Global governments, research laboratories, and other groups will continue their efforts to provide a renewable energy sustainable future. However, it will be the young people of today who are needed to champion the cause in order to reach the goal. Motivating them to reach the goal is the responsibility of their teachers, communities, mentors, peers, and parents.

Energy Data

Please note that energy data and statistics are constantly being revised by worldwide government agencies and nongovernmental organizations. However, the author has made a constant effort to include the most current data and statistics that were available to him at the time of publishing.

Chapter 1

Energy and Energy Sources

The Northeast Blackout of 1965 was a significant disruption in the supply of electricity on November 9, 1965, that left people without electricity for up to 13 hours in Ontario, Canada, and the U.S. states of New Hampshire, Vermont, Massachusetts, Rhode Island, Connecticut, New York, and New Jersey.

Paralyzing New York City in the middle of rush hour, the 13-hour blackout left 800,000 trapped in subways. People were isolated in elevators without lights or power. Airport lights on runways were out. The fire department and police officers had to set up temporary lights and barricades and stand at the intersections waving and directing cars and buses on their way. The blackout affected around 25 million people and stretched 80,000 square miles. This was the first time an enormous blackout occurred in the United States, and the events of the blackout help dramatize how much we depend on energy resources.

ENERGY'S ROLE IN OUR WORLD

Energy influences all aspects of our lives, and energy resources influence our current standard of living. Energy issues affect consumers, environmentalists, government leaders, investors, energy producers, and large and small businesses. Major energy news stories are reported each day in

In New York City, 42nd Street is lit by floodlights and automobile headlights during the massive power failure of November 9, 1965. The blackout affected New York State, most of New England, parts of New Jersey and Pennsylvania, and Ontario, Canada. (AP Photo)

magazines, newspapers, periodicals, newsletters, radio, and television and on the Internet.

Energy drives the global economy by producing much of the goods and services that are manufactured and sold in the marketplace. The global supply and demand for energy influences the major stock markets in all of the capitals of the world. Energy is at least a $7 trillion-per-year business, and it is expanding. Energy impacts all of our lives through its importance to transportation, electricity, manufactured goods, and agricultural production. As the historical blackout of 1965 proved, any disruption in an energy supply system or shortages of energy resources would have an impact on transportation, communications, national security, and economies in many countries.

ENERGY POLICIES NEED TO BE ADDRESSED

Presently, world governments, nongovernmental organizations, international companies, universities and colleges, entrepreneurs, and citizens are debating present and future energy policies. Following are a few of the questions they are addressing:

- By 2030 the world's population will have increased to more than 8 billion people. How will fossil fuel shortages, the depletion of non-renewable energy sources, and the rising costs of fuels, such as petroleum, impact the world's population of energy consumers?
- Will all proposed energy policies and programs be sustainable in format to meet the future energy needs and demands of people without damaging the environment?
- How reliable, efficient, and affordable are the renewable energy sources such as wind, solar, and hydrogen?
- What implementation plans are best to conserve energy usage in homes, businesses, transportation systems, and agricultural production?

There is no question that energy resources and technology will play a major role in global economics, public policy, science, ethics, and environmental issues as the 21st century continues.

HISTORY OF ENERGY IN THE UNITED STATES

In the United States, wood played a key role as an energy source in the early colonies until the mid-1880s. After this time period, coal replaced fuel wood in many states. Hydropower became another energy source in the 1930s, and by the 1950s petroleum had surpassed coal as the dominant energy source in the nation. The 1950s also saw the appearance of nuclear energy power plants. The most recent developments in energy resources include wind and wave energy, solar energy, geothermal energy, and biomass technologies.

U.S. Energy: To view U.S. energy use, go to www.teachersdomain.org/resource/tdc02.sci.life.eco.energyuse/.

U.S. HISTORICAL ENERGY DATA
Basic Energy Information

Date	Population	Production (in quads)	Consumption (in quads)
1950	151,326,000	35.6	34.6
1960	179,323,000	42.8	45.1
1970	203,302,000	63.5	67.8
1980	226,542,000	67.2	78.3
1990	248,422,000	70.7	84.6
2000	281,422,000	71.2	98.9
2006	299,338,000	71.0	99.9

Much of the energy demand in 2030 will still come from nonrenewable fossil fuels. These fuels will include the petroleum fuels, natural gas, and coal. (*Source:* U.S. Department of Energy/Energy Information Administration, *International Energy Outlook 2008*)

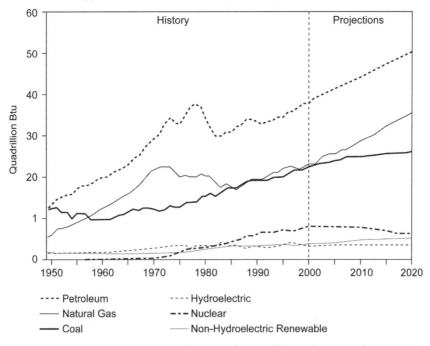

Nonrenewable energy sources like petroleum will continue to play a major role in energy consumption at least until 2020. (*Source:* U.S. Department of Energy/Energy Information Administration)

WHAT IS ENERGY?

Energy is the capacity for doing work or a force that produces an activity. Energy exists in many forms and can be converted from one form to another.

Forms of Energy

Energy can be discussed as either *potential* energy or *kinetic* energy. Simply stated, potential energy is the result of position and kinetic energy is energy of motion. Potential energy and kinetic energy are all around us and each has various forms.

Potential energy. There are several forms of potential energy, including:

Chemical energy. Chemical energy is stored in the bonds of atoms and molecules. Natural gas, petroleum, and coal are good examples of stored chemical energy.

Stored mechanical energy. Stored mechanical energy is the kind of energy that you would find in compressed springs in a grandfather clock or in a mechanical wristwatch.

Nuclear energy. Nuclear energy is stored in the nuclear structure of atoms. Energy can be released when the nuclei in atoms are split apart or combined or fused together.

Kinetic energy. The energy a substance or object possesses as a result of its motion is called kinetic energy. A moving train possesses kinetic energy. Kinetic energy is expressed as $MV^2/2$, a function of velocity (V) and mass (M). There are several forms of kinetic energy, including:

Radiant energy. Solar energy is an example of radiant energy. Radiant energy is electromagnetic energy that travels in transverse waves through empty space. Radio waves, visible light, and X-rays are examples of radiant energy.

 DID YOU KNOW?

The sun has produced energy for billions of years. In fact, all energy on Earth can be traced back to the sun except geothermal energy, which is derived from Earth's core.

- **Electromagnetic energy.** Electromagnetic energy consists of waves of electric and magnetic energy radiating through space and traveling at the speed of light.
- **Thermal energy or heat.** Geothermal energy is an example of thermal energy. The more atoms and molecules move in a gas, liquid, or solid, the more thermal energy there is in the material or substance.
- **Sound.** Sound is a form of kinetic energy in which molecules of air vibrate in a repeated pattern, causing the molecules to move in longitudinal wave patterns. Sound is produced from a force strong enough to make an object vibrate.
- **Motion.** Everything in the universe exhibits some form of motion. Wind is a good example of energy in motion. Motion is when objects move from one point to another.

And, of course, gravity is another source of energy, depending on position or the place of an object. Hydropower energy is a good example of gravity energy.

ENERGY CAN BE CONVERTED FROM ONE FORM TO ANOTHER

When energy is used, it does not disappear—it changes from one form of energy to another. For example, when natural gas is burned, it is converted to heat and light. This transfer of energy is based on the law of conservation of energy.

The Law of Conservation of Energy

The total quantity of energy available in the universe is a fixed amount, and there is never any more of it or less of it. Therefore, the law of conservation of energy (the first law of thermodynamics) states that energy cannot be created or destroyed, but energy can be transformed from one form to another—heat energy is transformed to light energy. As another example, when someone

 DID YOU KNOW?

People confuse the meanings of temperature and heat. Temperature is the measure of the average kinetic energy in a substance. Heat, on the other hand, is the total kinetic energy in a substance and can be measured with a thermometer.

strikes a match to ignite wood in a stove, the burning wood releases chemical energy that generates heat and light. A toaster uses electrical energy and converts it to thermal energy to toast food. Heat, light, and electricity are the most common byproducts of these conversions and the transfer of energy.

Multiple Conversions of Energy

Energy can also go through multiple transformations or conversions. Let's look at one example of a multiple conversion—wind energy. The motion of wind causes the mechanical energy of a wind turbine to spin a generator to produce electrical energy for consumers. The electricity can then be used in homes for thermal energy, light energy, and even mechanical energy to run a power tool.

Energy Loss in Conversions

Converting one form of energy into another form always involves a loss of usable energy that results in some of the energy changing into heat. This is the basis of the second law of thermodynamics. In most cases, heat from an energy conversion simply warms the surrounding air or solid material. The heat, which is not used to do work, is referred to as waste heat. Because of waste heat, some energy is lost in conversions; therefore, no machine is 100 percent efficient.

Cars and trucks jam a crowded freeway during the morning commute in Dallas, Texas. (iStockPhoto)

For example, today's gasoline internal combustion engines are not very efficient. The purpose of using the gasoline in the engine is to get the car moving. However, only about 15 percent of the chemical energy in the gasoline tank is used to power the motion of the car. What happens to the other 85 percent? It is wasted heat that escapes into the environment as the car moves along.

TODAY'S SOURCES OF ENERGY

According to the Department of Energy there are two major sources of energy that we use today. They include *nonrenewable* energy and *renewable* energy. Nonrenewable energy sources include oil and petroleum products, natural gas, coal, and nuclear energy. Renewable energy sources used most often are wind, solar, hydropower, geothermal, and biomass. The following is a short description of each of these sources of energy.

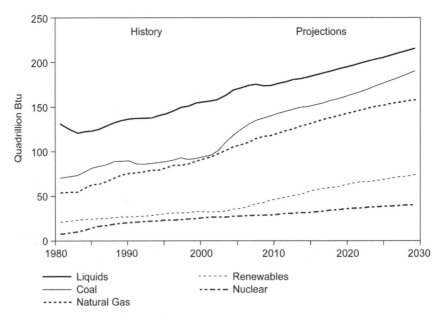

As you can see, the use of fossil fuels for energy will continue to grow into 2030. (*Source:* U.S. Department of Energy/Energy Information Administration, *International Energy Annual 2006* [June–December 2008]. Projections: EIA, World Energy Projections Plus [2009])

Nonrenewable Energy Sources

The United States and many of the other industrialized nations consume a large share of the world's total energy resources. The most common or conventional energy resources are fossil fuels. Fossil fuels are used to produce electricity and to operate automobiles, buses, trains, airplanes, and other machines. Approximately 85–90 percent of the energy consumed in the United States and the world comes from fossil fuels such as petroleum, natural gas, and coal.

Chapters 2, 3, 4, and 5 present more in-depth coverage of fossil fuels and nuclear energy. The following is a short description of each of these resources.

Fossil Fuels

Fossil fuels are naturally occurring nonrenewable energy sources, such as petroleum, natural gas, and coal. They are formed in Earth's crust over millions of years through the chemical and physical changes that occur in plant and animal remains. Large demands for fossil fuels began in the 18th century during the Industrial Revolution, and fossil fuels are a dominant energy source in the world today.

Petroleum resources. According to the International Energy Outlook 2009 Report, petroleum is expected to remain the world's dominant energy source through 2030. Presently, petroleum dominates the world's energy scene. Petroleum provides 40 percent of all of the energy in industrial countries and is the world's number one source of energy. The United States consumes approximately 26 percent of the world's production of petroleum.

Although gasoline is a major product of petroleum, other oil products manufactured include bottle gas (liquefied petroleum gas), kerosene, diesel fuel, asphalt, and plastic materials. Oil is especially critical for farmers, the transportation sector, and the chemical industry.

Natural gas. In the United States, natural gas ranks number three in energy use, right after petroleum and coal. Natural gas consists mainly of methane, the simplest hydrocarbon. About 22 percent of the energy we use in the United States comes from natural gas.

Today, residential and commercial uses account for the largest portion of natural gas consumed in North America and western Europe. In these areas, natural gas is commonly used for home heating and cooking. Gaseous fuels are convenient to use because they can be readily turned on and off, produce no smoke, and leave no ash behind.

After residential use, industry is the next-largest consumer, and electric-power generation is third. Major industries are the big consumers of natural gas, using it mainly as a heat source to manufacture goods and products, including fertilizers, paints, plastics, laundry detergent, and insect repellents. Many synthetic fibers such as those used in tires could not be made without the chemicals derived from natural gas. Natural gas can power vehicles, too. Some energy experts believe that the supplies of natural gas will be depleted by 2040.

Coal resources. According to a study by the U.S. Energy Information Agency (EIA), world consumption of coal is projected to increase from 123 quadrillion Btu in 2005 to 202 quadrillion Btu in 2030. The EIA also reports that by 2030 China will account for 71 percent of the world's consumption of coal. Worldwide, coal provides 40 percent of total electricity generation. The United States relies heavily on coal for electricity. Approximately 49 percent of U.S. electricity is from coal-powered generators. The top producers of coal in the world include China, the United States, India, Australia, and South Africa.

Today, coal's primary use is for the generation of electricity. However, coal is also used in the manufacturing of steel and cement.

Nuclear Energy

As of 2009 nuclear energy provided almost 20 percent of all electricity used in the United States and constituted about 15 percent of the world's electrical energy output according to the World Nuclear Industry Report.

The current conventional sources of electric power, such as coal, natural gas, and hydropower, may not be able to supply all of the world's electrical needs by 2030. The renewable energy sources such as wind, solar, and geothermal may still lag behind as major sources of electricity in the next 20 years. In fact, presently, the renewable non-hydropower fuels are used to meet less than 3 percent of electrical energy needs in the United States, according to the U.S. Energy Information Administration.

 DID YOU KNOW?

Coal mining in Pennsylvania fueled the Industrial Revolution in the United States in the mid-1700s.

On the global scene, as of 2009, 31 countries included nuclear power as part of their energy portfolio. These countries include the United States, Brazil, Egypt, China, Finland, India, Japan, Pakistan, Russia, Iran, South Korea, and Vietnam. According to the World Nuclear Association, as of 2007 there are 442 nuclear power reactors worldwide. These reactors supply approximately 15 percent of the world's electrical needs for more than 1 billion people without emitting any carbon dioxide or other greenhouse gases during their operation.

One of the major benefits of nuclear energy is that nuclear power plants can operate without contributing to climate change. Although the complete nuclear fuel cycle emits small amounts of greenhouse gases because of the fossil fuels used to mine uranium, transport nuclear fuel, and provide some of the electrical energy to run uranium enrichment plants, the amount of greenhouse gases emitted for the measure of electricity generated is lower for nuclear energy than for virtually all other electricity generation sources.

According to the World Nuclear Association, mainland China has 12 nuclear power reactors in operation and 24 under construction as of 2010. However, 80 percent of mainland China's electricity is produced from fossil fuels, mostly coal. (Shutterstock)

Renewable Energy Resources

Wind Energy

The fastest-growing renewable power source is wind energy. Wind energy, or wind power, is an alternative energy resource that uses the renewable energy in moving air to generate electricity. Although wind power currently produces less than 2 percent of the world's electricity, the Worldwatch Institute estimates that wind energy could easily provide 20–30 percent of the electricity needed by many countries. In the United States, the American Wind Energy Association (AWEA) estimates that by the year 2025, wind power will produce more than 10 percent of the electricity in the United States. Wind energy is discussed in more detail in volume 3.

Solar Energy

Solar energy is conversion of radiant energy from the sun into other forms of energy to provide solar heating and electricity. Presently, several technologies have been developed to use the sun's energy as a renewable energy resource for heat and electricity. The three key technologies include photovoltaic cells, concentrating solar power systems, and special solar collectors for space heating and hot water.

Photovoltaic (PV) cells convert sunlight directly into electricity. The cells are made of semiconductors such as crystalline silicon or various thin-film materials.

Concentrating solar power (CSP) technologies use reflective materials to concentrate the sun's heat energy. The heat eventually is used to drive a generator to produce electricity.

Low-temperature solar collectors, such as active and passive solar energy systems, absorb the sun's heat energy. The thermal energy is used directly for space heating or for hot water for homes and businesses.

Refer to volume 2 in this series for more solar energy information.

Hydroelectric Power

Hydroelectric power uses the kinetic energy of flowing water to drive wheels or turbines to generate electricity. The amount of electric energy produced by the generator depends on potential energy, which is dependent on the pressure and the volume of the water that flows into the turbine. Hydroelectric power accounts for about 22 percent of the world's electricity. Some of the largest hydroelectric power producers are Canada, the United States, Brazil, Norway, Russia, and China. Between 10 and 15 percent of all U.S. electricity is produced by hydropower.

Building small, rather than large, hydroelectric power systems may be the trend for the future. Today, small-scale hydroelectric power systems, called "mini-hydro" or "micro-hydro" systems, are being used on rivers and tributaries and in remote areas where construction is difficult. Such small-scale systems do not require the damming of rivers. These mini-hydro systems are used in China and the United States and in several smaller countries, including Indonesia, Nepal, Sri Lanka, and Zaire.

Refer to volume 3 in this series for more information about water power.

Geothermal Energy and Heat Pumps

Geothermal energy refers to the use of natural heat energy that is extracted from the interior of Earth in the form of steam, hot water, and hot dry rocks. Geothermal energy is an alternative energy resource that can be used for the direct heating of buildings or for generating electricity. Geothermal energy is not always listed as a renewable energy source because in some locations the depletion rate of sources such as hot water can be higher than the rate at which the sources replenish or recharge. Italy, Iceland, New Zealand, Russia, Japan, and France, along with the United States, are countries using geothermal energy. Other countries using geothermal energy include the Philippines, Indonesia, Mexico, countries in Central and South America, and countries in eastern Africa and in eastern Europe.

Refer to volume 4 in this series for more geothermal energy information.

Biomass

The energy from biomass is the oldest fuel used by humans, and fuelwood is the most widely used biomass fuel. However, there are other sources for biomass energy, including herbaceous plants and excess food crops that can be burned as a direct source of energy. Unused parts of sugar cane, cornstalks, peat, and even cattle dung have been used as biomass fuels. Even municipal solid wastes, a form of biomass, can be burned directly as fuel. In Europe processing plants use up to 50 percent of municipal trash for energy production. Trash-to-energy plants are also located in several American cities in Maryland, California, Illinois, Ohio, Wisconsin, and Washington.

Role of Renewable Energy in Nation's Energy Supply (2007)

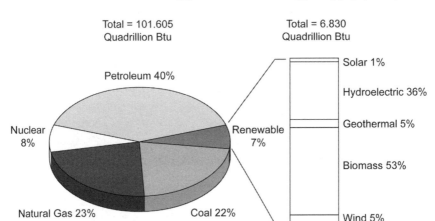

As of 2007, less than 10 percent of the United States' supply of energy was derived from renewable energy sources. Petroleum was by far the major source of energy for the country. (*Source:* U.S. Department of Energy/Energy Information Administration, *Renewable Energy Consumption and Electricity Preliminary 2007 Statistics. Note:* The sum of the components may not equal 100 percent due to independent rounding.)

Fuelwood. As noted previously, fuelwood is the most widely used biomass fuel. For thousands of years, human populations have harvested wood as the most common source of energy. Today, fuelwood is used as a common source of energy for cooking and heating by as many as 3 billion people throughout the world. In fact, only in the last few hundred years, since the Industrial Revolution, have people started using other sources of energy, such as fossil fuels.

Biofuels. Biofuels are solid, liquid, or gaseous fuels derived from biomass sources, which contain stored energy. Biofuels are used as an alternative to fossil fuels and include biogas, biodiesel, and methane. About 5 percent of the energy consumed in the United States is provided by biofuels. Most of the biofuels are produced from wood waste from logging operations, but they can also be produced from corn and sugar crops. In France, Italy, and Germany, biodiesel fuels are produced from domestic oilseeds and cottonseeds. Biofuels are cleaner than fossil fuels because they release few greenhouse gases, sulfur, and particulate matter into the atmosphere.

Refer to volume 4 in this series for more information about biomass and biofuel energy.

> **DID YOU KNOW?**
>
> The International Energy Agency reports that less than a third of the households in many oil- and gas-rich nations have access to electricity or clean fuels for cooking. It is estimated that approximately 150,000 people, mainly women and children, die prematurely each year because of indoor air pollution from burning traditional fuels such as fuelwood and charcoal in inefficient stoves or open fires. The number of deaths will rise as population grows.

Ocean Thermal Energy Conversion (OTEC)

Ocean Thermal Energy Conversion (OTEC) is an alternative energy resource that uses the natural temperature differences between various layers of ocean water to produce electricity. The idea of using OTEC to produce electricity is not new. A small OTEC plant was built off the coast of Cuba in the 1930s. The plant produced electricity for the island country until it was destroyed by a hurricane. Another plant was built in 1956 off the coast of Africa. Later, a dam that generates electricity via hydroelectric power at a lower cost replaced this plant.

OTEC systems work best in the tropical waters of the central Pacific Ocean and the Indian Ocean and in the Gulf of Mexico region of the Atlantic Ocean.

Refer to volume 4 in this series for more information on hydropower, ocean tidal power, ocean wave power, and ocean thermal energy.

Hydrogen Fuel Cells

Some energy consultants state that someday a hydrogen fuel cell will be used to produce electricity to power automobiles, machines, and even homes. Hydrogen is the lightest and most common element in the world. Today, hydrogen is used primarily in ammonia manufacturing and petroleum refining.

Hydrogen fuel cells are also used by NASA, which has installed fuel cells aboard the space shuttles. The fuel cells provided heat, electricity, and drinking water for the astronauts. The good news is that when hydrogen is used as an energy source, it generates no emissions other than water, which can be recycled to make more hydrogen.

Refer to volume 2 in this series for more information about hydrogen fuel cells.

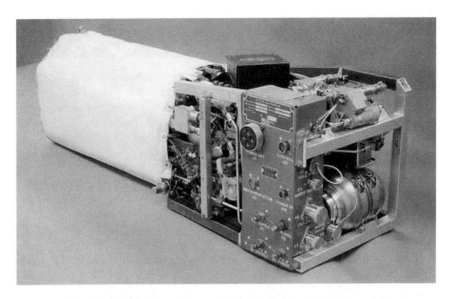

The National Aeronautics and Space Administration (NASA) uses hydrogen fuel cells in its space shuttle program. Someday, fuel cells the size of a refrigerator may be used to provide electricity and heat for homes and other buildings. (National Aeronautics and Space Administration)

ELECTRICITY, A SECONDARY SOURCE OF ENERGY

Electricity is named as a secondary source of energy because it is generated by other energy sources such as petroleum, wind, solar power, coal, or natural gas.

What Sources of Energy Are Used to Produce Electricity?

In the United States, the three kinds of power plants that produce most of the electricity are those using fossil fuels, nuclear energy, and hydropower. Coal power plants produce almost 50 percent of all electricity used in the United States. Solar power plants, wind farms, geothermal plants, and the burning of biomass such as trash for electrical power generate only about 2 percent of all electricity.

The thermal power plants use coal, petroleum, and natural gas to convert water into steam. The steam is pumped through a pipe at high pressures to spin a generator, which makes the electricity. The nuclear power plant uses

Energy and Energy Sources 17

View of the massive Hoover Dam from a helicopter. Originally known as Boulder Dam, Hoover Dam is located on the Nevada–Arizona border in the Black Canyon of the Colorado River. (iStockPhoto)

fission to produce the heat. The hydropower plants use fast-moving water to spin the blades of a generator to produce electricity.

How Is Electricity Transmitted to Homes and Businesses?

The electric utility company uses generators to produce much of the electricity we use today in homes, businesses, schools, and farms. For electricity to reach all of these places, the electrical energy needs to be transmitted over long distances at very high voltages, between 11,000 volts and 700,000 volts. Special step-up transformers are used to transmit the high-voltage electrical energy along transmission lines from the power plants.

Special step-down transformers are used to decrease the voltage, so that the electricity can be used safely in our homes and schools. The high voltage is eventually reduced to 220 volts for appliances such as stoves and clothes dryers and to about 120 volts for lights and other smaller appliances.

 DID YOU KNOW?

Step-up transformers are installed in television sets because they require higher voltages than the current in your home.

HOW IS ENERGY MEASURED?

Energy can be measured using a variety of units. The amount of heat consumed or generated from various types of energy can be measured by both the calorie unit and the British thermal unit (Btu).

Calorie. A calorie is a measure of heat energy. Using the International System of Units (SI) for measurement, a calorie is the amount of heat energy needed to raise the temperature of one gram of water by one degree Celsius.

British thermal unit (Btu). The Btu is also commonly used to measure the amount of heat consumed or generated from various types of energy. The English unit of measurement defines the Btu as the quantity of heat required to raise the temperature of one pound of water one degree Fahrenheit at a normal atmospheric pressure. One Btu is the energy equivalent of one burning match tip; or, 1 Btu equals 252 calories or 1,055 joules.

Each gallon of petroleum produces about 12,500 Btu. One cubic foot of natural gas produces between 900 and 1,200 Btu of energy. One pound of bituminous coal is equal to 12,000 Btu. Propane has a heating value of 2,500 Btu per cubic foot. About 100 cubic feet of natural gas would contain 100,000 Btu and is referred to as a "therm."

Kilowatt-hour (kWh). This is a common unit in which electricity is measured. It is equal to 1,000 watts or 1,000 joules per second. A kilowatt-hour (kWh) is a unit of electrical energy equal to 1,000 watt-hours. One kWh is equal to 3,413 Btu. Utility companies bill their customers in cents per kilowatt-hour. The average home in the United States uses about 9,400 kWh of electricity a year.

Quad. A quad is a unit of measurement equal to 1 quadrillion Btu, or 1,000,000,000,000,000 Btu (or 10^{15} Btu). Scientists measure large quantities of energy using quad measurements. The United States averages one quad of energy about every 3.7 days.

Cubic foot. A cubic foot is a measure of volume. One cubic foot of natural gas contains about 1,020 Btu. One thousand cubic feet of natural gas for residential consumers costs between $10 and $15.

WORLDWIDE CONSUMPTION OF ENERGY

The world population is expected to grow to 10 billion people by the middle of the 21st century. At the same pace, it is expected that worldwide consumption of energy will have increased by 50 percent in 2030,

U.S. Energy Consumption by Energy Source, 2005 - 2009
(Quadrillion Btu)

Energy Source[1]	2005	2006	2007	2008	2009
Total	100.468	99.790	101.502	99.438	94.820
Fossil Fuels	85.815	84.687	86.223	83.532	78.631
Coal	22.797	22.447	22.749	22.398	19.996
Coal Coke Net Imports	0.045	0.061	0.025	0.040	-0.023
Natural Gas[2]	22.583	22.224	23.679	23.814	23.416
Petroleum[3]	40.391	39.955	39.769	37.279	35.242
Electricity Net Imports	0.084	0.063	0.106	0.113	0.116
Nuclear Electric Power	8.161	8.215	8.455	8.427	8.328
Renewable Energy	6.407	6.825	6.719	7.367	7.745
Biomass[4]	3.117	3.277	3.503	3.852	3.884
Biofuels	0.577	0.771	0.991	1.372	1.546
Waste	0.403	0.397	0.413	0.436	0.447
Wood and Derived Fuels	2.136	2.109	2.098	2.044	1.891
Geothermal Energy	0.343	0.343	0.349	0.360	0.373
Hydroelectric Conventional	2.703	2.869	2.446	2.512	2.682
Solar Thermal/PV Energy	0.066	0.072	0.081	0.097	0.109
Wind Energy	0.178	0.264	0.341	0.546	0.697

[1] Biodiesel primarily derived from soybean oil and ethanol primarily derived from corn.
[2] Includes supplemental gaseous fuels.
[3] Petroleum products supplied, including natural gas plant liquids and crude oil burned as fuel.
[4] Biomass includes: biofuels, waste (landfill gas, MSW biogenic, and other biomass), wood and wood derived fuels.
PV = Photovoltaic.
Notes: Data revisions are discussed in the Highlights section.
Totals may not equal sum of components due to independent rounding.
Data for 2009 is preliminary.

U.S. energy consumption by quad. A quadrillion is equal to 1 trillion megawatts (MW). (*Source:* U.S. Department of Energy/Energy Information Administration/Renewable Energy Consumption and Electricity Preliminary Statistics 2009)

in comparison with 2005 consumption. Through 2030, fossil fuels such as petroleum and coal are expected to continue to supply much of the energy used worldwide, according to the International Energy Outlook report, published by the U.S. Department of Energy in 2008. World energy consumption is projected to increase 40–50 percent from 2005 to 2030. The following resources will produce this energy:

- Fossil fuels such as coal and petroleum (78%)
- Renewables such as hydropower and wind (18%)
- Nuclear (4%)

WORLDWIDE USES OF ENERGY

The growing world population will continue using energy for many reasons, including for transportation, electricity, heating needs, and industry.

According to the EIA, as of 2007 the industrial sector uses more energy globally than the transportation and building sectors. The industrial

World Marketed Energy Consumption 2006-2030

Quadrillion Btu:
- 2006: 472
- 2010: 508
- 2015: 552
- 2020: 596
- 2025: 637
- 2030: 678

(Legend: Non-OECD, OECD)

Members of the Organization of Economic Cooperation and Development (OECD) include the United States, Japan, Australia, Turkey, Chile, and many of the European countries. Two non-OECD countries are Brazil and China. (*Source:* U.S. Department of Energy/Energy Information Administration, *International Energy Annual 2006* [June–December 2008]. Projections: EIA, World Projections Plus [2009])

sector consumes 50 percent of the world's total energy. This sector includes manufacturing, agriculture, mining, and construction. Some of the major energy-intensive industries according to their statistics, include the following:

Petroleum. Oil refineries use a lot of energy to convert crude oil into a variety of products, including gasoline, diesel fuel, heating fuel, and chemicals. In fact, 50 percent of a refinery's operating costs are for energy.

Steel manufacturing. The steel industry uses energy to produce steel for hundreds of products. Steel is made from iron ore and other materials at very high temperatures. Steel products include home appliances and automobile parts. Producing these high temperatures with coal-fired furnaces is very costly; however, 66 percent of new steel is made from recycled scraps, making steel the leading recycled product in the United States.

Aluminum manufacturing. The manufacture of aluminum, like steel production, requires large amounts of energy to produce a variety of products, including beverage containers, food trays, and automobile parts.

Paper manufacturing. The manufacture of paper products includes a number of steps such as chopping, grinding, and cooking the wood or recycled materials into pulp. All of these steps require energy.

Other large industrial energy users include the chemical manufacturing industries and cement manufacturing companies. These industries use coal, oil, and natural gas to produce the energy needed for high-temperature manufacturing processes.

The transportation sector uses energy that is consumed for moving people, goods, and fuels, using a variety of transportation systems that include trucks, cars, buses, subways, railroads, ships and barges, airplanes, and pipelines. Almost 30 percent of the world's total energy is used for transportation, and most of that energy is used in the form of liquid fuels.

The building sector, which consists of homes and commercial buildings, accounts for about 20 percent of the world's total energy consumption. This energy is used for heating, lighting, air conditioning, and for powering appliances used for cooking, refrigeration, and entertainment systems. As of 2007, 40–50 percent of all global electricity production is generated at coal-fired power plants. Natural gas is also used for heating and hot water needs for homes and businesses.

FUTURE WORLD ENERGY CONSUMPTION

As of 2009, according to government reports and energy studies, petroleum, natural gas, and coal are expected to remain the world's dominant energy source throughout the next 20 years. The reason is that these fossil

 DID YOU KNOW?

As of 2007 there are, worldwide, more than 700 million cars and other vehicles on the road.

fuels will still be the major sources of energy for transportation and for industrial production of goods and products.

According to a report by Europe's Energy Portal Organization, global energy consumption is projected to grow by 44 percent from 2006 to 2030. Total world energy use rises from 472 quadrillion British thermal units (Btu) in 2006 to 552 quadrillion Btu in 2015 and then to 678 quadrillion Btu in 2030. In 1980 the world consumption was approximately 283 quadrillion Btu. If this is true, the world's energy use will increase by 150 percent in less than 50 years.

Environmental Concerns and Implications for Climate Change

The rising consumption of fossil fuels will increase the world greenhouse gas emissions and global temperatures. These conditions will cause the potential for major climate change that may not be reversible.

Fossil Fuel Emission Levels
-Pounds per Billion Btu of Energy Input

Pollutant	Natural Gas	Oil	Coal
Carbon Dioxide	117,000	164,000	208,000
Carbon Monoxide	40	33	208
Nitrogen Oxides	92	448	457
Sulfur Dioxide	1	1,122	2,591
Particulates	7	84	2,744
Mercury	0.000	0.007	0.016

Carbon dioxide is the major pollutant in fossil fuel emissions. (*Source:* U.S. Department of Energy/Energy Information Administration, *Natural Gas 1998: Issues and Trends*)

DID YOU KNOW?

Among industrialized and developing countries, Canada consumes per capita the most energy in the world; the United Sates ranks second, and Italy consumes the least among industrialized countries.

Carbon Dioxide and Greenhouse Gases

Many gases exhibit so-called greenhouse properties. Such gases absorb infrared radiation, trapping it within Earth's atmosphere. Some greenhouse gases occur naturally, such as carbon dioxide (CO_2), methane, water vapor, and nitrous oxides. Others are produced exclusively by manufacturing activities. They include chlorofluorocarbons, hydrofluorocarbons, and perfluorocarbons.

However, the major greenhouse gas is carbon dioxide. Carbon dioxide is a colorless odorless gas that plays a key role in controlling temperatures at Earth's surface. Carbon dioxide contributes to the greenhouse effect much as the glass walls of a greenhouse trap heat, and that is good because it keeps Earth warm so that humans and other organisms can survive. But too much CO_2 building up in the atmosphere can be a problem.

As global energy consumption increases, so will CO_2 emissions. According to some environmental reports, CO_2 emissions are projected to rise from 30 billion tons in 2006 to 36.1 billion tons in 2015 and 44 billion metric tons in 2030—an increase of 39 percent over this projection period, if the data is correct.

"The International Energy Outlook 2006 foresees a rise of global energy consumption by 71% between 2003 and 2030, resulting in an increase of world-wide carbon dioxide emissions by 75% over the same period." Approximately 75 percent of the projected increase in emissions will be in China, India, and the Middle East. Only in Europe and in Japan will the emissions be lower in 2030 than they were in 2003.

In the future, more and more carbon dioxide will be released into the atmosphere than ever before and thus could contribute to a much warmer Earth in the future. Some studies of carbon dioxide levels in the atmosphere support this view. As atmospheric temperatures rise, global temperatures also rise, causing global warming.

Global Warming

Global warming is a recent, ongoing elevation in global surface air temperature primarily resulting from human-caused increases in the concentrations of greenhouse gases such as carbon dioxide in the lower atmosphere. Carbon dioxide is produced when fossil fuels are used to generate energy and when forests are cut down and burned. It is estimated that 76 percent of global warming is caused by carbon dioxide alone. The average concentration of carbon dioxide increased from about 275 parts per million (ppm) before the Industrial Revolution to 315 ppm when precise monitoring stations were set up in 1958 to 361 ppm in 1996. If increased concentrations of carbon dioxide get to approximately 1,000 parts per million of carbon dioxide, an eventual global temperature increase of up to 12 degrees Fahrenheit would result.

Global average temperatures have remained relatively stable over the last 10,000 years. But since 1880, when reliable temperature records started to be kept worldwide, the global average temperature has risen by nearly 15 degrees Fahrenheit. Snow cover in the Northern Hemisphere and floating ice in the Arctic Ocean have decreased, and cold-season precipitation has increased in the high latitudes. According to some reports, globally, sea level has risen 4–10 inches over the past century. Earth's northern latitudes have become much greener during the growing seasons since 1980, and the spurt in plant growth may be associated with warmer temperatures and higher levels of atmospheric carbon dioxide, which plants take in.

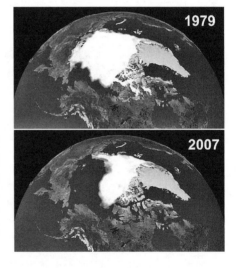

These satellite images were used to compare Arctic ice mass in 1979 and 2007. Notice the differences in the ice mass in the western section of the Arctic over this 28-year period. (National Aeronautics and Space Administration)

VIDEO

Energy versus Fossil Fuels: This 02:09 minute video lays the foundation for fossil fuels and their impact on global warming. For more, go to http://videos.howstuffworks.com/hsw/6189-energy-fossil-fuels-video.htm.

What Are Some of the Plans to Reduce Climate Change and Global Warming?

Tracking the climate change and global warming issue and at the same time providing energy resources for a growing global population will require some innovative solutions. Some of these plans include curbing carbon emissions, investing more research in renewable energy technologies, and promoting energy efficiency and energy conservation programs.

Saving Resources and Saving Energy

There are many ways to save energy and conserve our natural resources via energy conservation and energy efficiency. Many people think these terms mean the same thing, but they are different. The Department of Energy describes the distinction in these terms:

> The terms energy conservation and energy efficiency have two distinct definitions. There are many things we can do to use less energy (conservation) and use it more wisely (efficiency).
> - **Energy Conservation** is any behavior that results in the use of less energy. Turning the lights off when you leave the room and recycling aluminum cans are both ways of conserving energy.
> - **Energy Efficiency** is the use of technology that requires less energy to perform the same function. A compact fluorescent light bulb that uses less energy than an incandescent bulb to produce the same amount of light is an example of energy efficiency. However, the decision to replace an incandescent light bulb with a compact fluorescent is an act of energy conservation.

The Department of Energy report goes on to show how recycling functions as a type of energy conservation:

> Recycling means to use something again. Newspapers can be used to make new newspapers. Aluminum cans be used to make new aluminum cans. Glass jars can be used to make new glass jars. Recycling often saves energy and natural resources.
>
> Natural resources are things of value provided by the Earth. Natural resources include land, plants, minerals, and water. By using materials more than once, we conserve natural resources....

It almost always takes less energy to make a product from recycled materials than it does to make it from new materials. Using recycled aluminum scrap to make new aluminum cans, for example, uses 95% less energy than making aluminum cans from bauxite ore, the raw material used to make aluminum.

In the case of paper, recycling saves trees and water. Making a ton of paper from recycled paper saves up to 17 trees and uses 50% less water.

Schools Becoming Energy Smart

Schools can play a leadership role in energy conservation and efficiency. In addition to the benefits school conservation programs provide to the environment, some schools have reported savings as high as 25 percent on their utility bills.

Among the strategies schools have employed in meeting emissions-reductions targets are introducing passive heating and cooling techniques, converting to renewable energy sources, and retrofitting buildings for energy efficiency. Other programs address such practices as composting and recycling. The most successful programs involve all members of the school

In Chicago, a living, green roof was planted atop the gymnasium at the Tarkington Elementary School. The insulation provided by the soil and vegetation will help keep the building warm in winter and cool in summer. (AP Photo/ Nam Y. Huh)

community, including students, staff, teachers, and administrators, all of whom play an important role.

One school in Maine, as an example, uses technology to provide computer-controlled bank lighting in classrooms. When the sunlight is bright, the lights are dimmed automatically. Occupancy (motion) sensors in classrooms also turn lights off automatically when the classrooms are not in use. These lighting strategies help to reduce electricity consumption and save money.

Another school in Maine reported saving over 9,000 gallons of heating fuel between September and December 2007 after installing a new burner control system. The new system has auto flame control with variable frequency drive, which uses the most efficient mix of air and fuel for the amount of ventilation needed at the time. The school has saved over $20,000 and prevented over 200,000 pounds of carbon dioxide emissions. The Maine schools were assisted by the Department of Environmental Protection workshops and programs.

The Alliance to Save Energy. There are several government agencies and nongovernmental organizations that assist schools in becoming more "green" by saving energy and reducing carbon dioxide emissions. One organization called Alliance to Save Energy has a special program called Alliance's Green Schools Program. The associates are businesses and nonprofit organizations committed to greater investment in energy efficiency as a primary means of achieving the nation's environmental, economic, national security, and affordable housing goals.

According to the organization's Web site: "Founded in 1977, the Alliance to Save Energy is a non-profit coalition of business, government, environmental and consumer leaders. The Alliance to Save Energy supports energy efficiency as a cost-effective energy resource. It also advocates energy-efficiency policies that minimize costs to society and individual consumers, and that lessen greenhouse gas emissions and their impact on the global climate."

The Alliance's Green Schools Program engages students in creating energy-saving activities in their schools. The students participate in hands-on, real-world projects, and Green Schools has achieved reductions in energy use of 5–15 percent among participating schools. Past student participants in the Green Schools Program have made presentations to school boards on energy-efficiency retrofit recommendations, authored pieces for the local newspaper, and conducted energy audits for local small businesses, among other activities. Several school districts participate in the program.

In the 2008–2009 school year, the Alliance's Student Energy Audit Training (SEAT) program taught students in the District of Columbia. The training sessions included students using auditing tools such as light meters, watt meters, and infrared thermometers to detect areas of energy waste in their schools. The students calculate savings from conservation activities such as turning off lights, removing bulbs from fixtures, and switching to more efficient bulbs.

To learn more about the Alliance to Save Energy's Green Schools Program, visit http://ase.org/programs/green-schools-program.

Refer to volume 5 in this set for more information on energy efficiency, energy conservation, and energy sustainability.

 INTERVIEW

Green Advocate: Linda Currie (Linda@baylocalize.org), Energy Outreach Coordinator, Bay Localize, www.baylocalize.org, Oakland, California
Go-Green Project: Energy-Efficiency Methods in Homes

Describe your current position and your responsibilities in your particular program.

I work for Bay Localize (www.baylocalize.org), a nonprofit whose goal is to reduce reliance on fossil fuels while increasing community resilience and livability in the nine County San Francisco Bay Areas.

As energy outreach coordinator for the "Green Faith in Action Project," I am working with church congregations on home energy-efficiency upgrades performed by graduates of several green jobs training programs.

Where did you grow up, and what schools did you attend?

I grew up in the Midwest, mostly in Indiana and Ohio. My dad was an industrial engineer and always rigging up systems around the house to save energy, one of which was a device that would prevent the heating thermostat from being set above 60 degrees Fahrenheit in the winter. My mom was a dietitian, and from an early age, I learned the downfalls of fast food.

In seventh grade, my Girl Scout troop took a long bus trip throughout the Southwest. We hiked to the bottom of the Grand Canyon and stayed at the Phantom Ranch by the Colorado River. I will never forget the experiences we had with nature and the surreal beauty of places like Bryce and Zion Canyons. In scouting we had a rule about leaving places we stayed in better shape than we found them, a pledge I try to adhere to in my life and work today. My family camped in many places around the country including Florida, Maine, and California. The more nature I saw, the more I loved it and wanted to help preserve it.

What were some of your favorite activities and subjects when you attended high school?

I loved many subjects in school, especially art and science, and had several great science teachers in high school. One of my favorite classes was anatomy, where we got to dissect cats.

What college(s) did you attend, and what was your major field?

I went on to attend Bowling Green State University, where I took a variety of courses including biology classes and was particularly fascinated by marine biology and the color and diversity of sea life. I studied art as well and graduated with a bachelor of fine arts degree, with a specialty in graphic design.

What interested you in seeking a career in your discipline?

I have a lot of varied interests and started my career in the graphic design field, where I spent many years working as an art director for the University of California at Berkeley, in the public affairs department. Some of my projects involved highlighting the groundbreaking research projects in the sciences and other disciplines.

In more recent years, as someone who deeply appreciates our interconnectedness with nature, I decided to take an active role in helping preserve our planet. I felt I needed to do whatever I could to help mitigate climate change, so my children and future generations could enjoy a safe and healthy environment. I had always been community-minded, serving on many school committees, so community organizing seemed the natural direction for me. I started learning as much as I could about climate change by going to meetings hosted by the city of Berkeley when they began to develop their climate action plan in 2007. I also joined an organization called Green Sangha (www.greensangha.org), where I met others seeking a positive way to take action on environmental issues.

One action some members of our group took was to go through the "Low Carbon Diet" (www.empowermentinstitute.net), a four-session program developed by David Gershon to analyze your carbon footprint and then take simple steps to reduce it. The program was so successful for members of our group and for my family that I began to form and facilitate other groups within my community. I found that places such as schools and churches, with natural community affiliations, were great places to help people make changes in living more sustainably.

Describe the Green Faith in Action Project.

The work I'm doing with the Green Faith in Action Project involves partnering with Rising Sun Energy Center and its CYES (California Youth Energy Services) program. CYES trains young people (15–22 years of age) to conduct free "green house calls" to install energy-efficiency measures during the summer months. At each "green house call" the youth energy specialist educates the resident about sustainability, installs free equipment that saves energy and water, and provides personalized recommendations for further energy savings.

We are also working with Richmond BUILD's GETS (Green Jobs Training) program, using recent graduates to perform home energy-performance tests and more advanced home energy-efficiency upgrades. Richmond BUILD (Pre-apprenticeship

California Youth Energy Services (CYES) youth energy specialists Alexandria Parr and Jordan Flores install a compact fluorescent lightbulb during a "green house call." (Courtesy Rising Sun Energy Center [www.risingsunenergy.org])

Construction Skills & Green Jobs Training Academy was first developed to create employment and career opportunities for Richmond residents and also to implement a strategy for reducing violence. Richmond BUILD was established in April 2007 and has quickly become a model of effective and broad public–private partnership that is focused on developing talent and skills in the high-wage construction and renewable-energy fields.

Our Green Faith in Action Project will connect the green jobs trainees with faith communities in Marin and Richmond, California, to have their homes made more energy-efficient. In order to measure the impact of the program, we will be closely monitoring and analyzing participant energy use.

Following are outcomes we hope to achieve:

- Make participants' homes more comfortable in both hot and cold weather
- Help participants save money on energy bills
- Allow participants to take action on climate change and clean up our air
- Give youth and new trainees hands-on experience to launch their green careers
- Measure the impact of how much money and energy participants have saved collectively

Even in its initial stages, this program is already being replicated by the City of Berkeley, and was implemented in 2010. Thanks go to the Frank

Levinson Fund at the Silicon Valley Community Foundation for sponsoring this program.

How or why did you get interested in selecting this project?

I had done some work with church congregants in helping them learn to live more sustainably and saw this project as another avenue of continuing that work with the potential of creating a model for other faith-based groups to follow.

Explain the importance of the project as it relates to real-world issues.

Residential energy-efficiency audits and upgrades are a cost-effective strategy to reduce energy use, save money for households, help municipalities reach climate action goals, and promote community development by providing local jobs.

What materials and references (web sites, advisors, periodicals, etc.) did you use as resources for the project?

Rising Sun Energy Services: www.risingsunenergy.org/cyes.htm

Richmond Build: www.ci.richmond.ca.us/index.aspx?nid=1243

Interfaith Power and Light: www.theregenerationproject.org/

Energy Star for Congregations: www.energystar.gov/index.cfm?c=small_business.sb_congregations

PG&E (Pacific Gas and Electric) Energy Savings Tips: www.pge.com/myhome/saveenergymoney/savingstips/index.shtml

PG&E Rebates: www.pge.com/myhome/saveenergymoney/rebates/

LIHEAP (Low Income Home Energy Assistance Program): www.acf.hhs.gov/programs/ocs/liheap/

U.S. Department of Energy: www.energy.gov/energyefficiency/index.htm

What advice would you give other teachers who would like to know more about your program or activity?

Feel free to contact me by e-mail: Linda@baylocalize.org.

Discuss some of the students' contributions to this project, and describe some of their reactions or comments.

Students in both the CYES and GETS programs are getting valuable training and skills that are necessary in California for meeting its climate change goals of making homes more energy-efficient. The California Public Utilities Commission (CPUC) has established a goal of making every California home up to 40 percent more efficient by the year 2020. Accomplishing this task will require a massive, coordinated effort of outreach and marketing to homeowners, offering financing options, and training and certifying a workforce to carry out the contracts.

How long did it take to complete the project?

This project is in the beginning stages.

Do you have any current plans to improve on or to extend this program? If not, what are you planning next in the field of energy, conservation, or the environment?

We'll wait to see the results before planning the next steps.

Global Plans to Reduce Emissions

1997 Kyoto Protocol

For many years, nations have been working on plans to establish global regulations to reduce overall emissions. The 1997 Kyoto Protocol, which authorizes the curbing of emissions, was signed by 182 nations to meet the goals for reducing emissions of greenhouse gases, such as carbon dioxide. Although the United States did not sign the Kyoto Protocol, many cities and towns throughout the United States are already moving forward with carbon-cutting plans on their own. In fact, several states are requiring electricity producers to reduce carbon emissions by 10 percent by 2018. Other state governments are requiring that a certain percentage of their electricity be produced from renewable energy sources such as wind and solar.

Countries such as Germany and China are replacing some of their fossil fuel plants with hydropower and other renewables. Germany is planning to

Kyoto governor Teiichi Aramaki makes a speech during the opening session of the United Nations Framework Convention on Climate Change on December 1, 1997, in Kyoto, Japan. The world's nations convened for an extraordinary 10 days of negotiations about the Earth's future. (AP Photo/Katsumi Kasahara)

remove all of its coal-producing plants in the near future. Many countries in Europe and Asia are installing solar-powered, wind-powered, and geothermal energy systems to reduce their need for fossil fuels.

The United States is also working on plans to use more renewable energy sources.

U.S. Economic Stimulus Bill, 2009

In 2009 the U.S. government passed a stimulus package, the American Recovery and Reinvestment Act, to jump-start the economy through job creation and with a heavy focus on energy. According to the Department of Energy, following are several provisions in the stimulus package:

- Tax credits for the production of renewable energy are extended until at least 2012.
- Research expenses associated with renewables, conservation, and carbon capture and sequestration could result in higher credits in both 2009 and 2010.
- The Department of Energy is authorized to provide grants up to 30 percent of the cost of installation of items such as fuel cells, solar and small wind power, geothermal heat pumps, and combined heat and power systems.
- The Department of Energy's Office of Energy Efficiency and Renewable Energy was to receive $21.4 billion for research, weatherization assistance, grants, and other programs.
- The Department of Labor was to receive $750 million for job training, with significant focus on emerging industry sectors, including energy efficiency and renewable energy.
- Federal agencies were to receive considerable funds for retrofitting and upgrading existing facilities to meet federal energy and water-use requirements and alleviate any maintenance backlogs.

"In 2009, the White House announced that the U.S. Department of Energy Office of Science will invest $777 million in Energy Frontier Research Centers [EFRC] over the next five years. The EFRCs will bring together groups of leading scientists to address fundamental issues in fields ranging from solar energy and electricity storage to materials sciences, biofuels, advanced nuclear systems, and carbon capture."

FEATURE

U.S. Department of Energy

A department of the federal government established in 1977 to regulate and manage the energy policy of the United States, the Department of Energy (DOE) consolidated many of the federal government's responsibilities for energy and national defense materials into one agency by replacing earlier energy-related agencies of the federal government.

The main responsibilities of the DOE include providing technologies and developing policies that achieve efficiency in energy use while maintaining environmental quality and a secure national defense. The department is a world leader in the research and development of programs and technologies that generate energy from fossil fuels, nuclear fuels, and alternative energy resources such as solar energy, wind power, and biofuels. In addition, the DOE administers comprehensive environmental management programs involving the cleanup of sites contaminated with high-level radioactive wastes (HLRW) and other contaminants resulting from energy generation. The DOE also manages the nation's hydroelectric power plants, such as the Bonneville Power Administration of the northwestern United States.

BOOKS AND OTHER READING MATERIALS

Berinstein, Paula. *Alternative Energy: Facts, Statistics, and Issues.* Phoenix: Oryx Press, 2001.

Boyle, Godfrey, ed. *Renewable Energy.* Oxford, UK: Oxford University Press, 2004.

Chandler, Gary, and Kevin Graham. *Alternative Energy Sources (Making a Better World).* New York: Twenty First Century Books, 1996.

Graham, Ian. *Fossil Fuels: A Resource Our World Depends Upon.* Chicago: Heinemann Library, 2005.

Richard, Julie. *Fossil Fuels.* North Mankato, MN: Smart Apple Media, 2003.

SOMETHING TO DO

Energy has been defined as the ability to do work or change. We cannot see energy, but we can tell if it is there. Observe a particular environment and complete a data sheet that answers the following questions about nonrenewable energy sources in that area.

What nonrenewable energy sources do you observe being used? How are they being used? Who uses them? How are they produced and distributed? What are the advantages and disadvantages of each? How might they be used more efficiently?

WEB SITES

The following Web sites, though not inclusive, include government and nongovernmental organizations.

www.americancoalcouncil.org
> The American Coal Council (ACC) is dedicated to advancing the development and utilization of American coal as an economic, abundant/secure, and environmentally sound energy fuel source.

www.aga.org
> The American Gas Association, founded in 1918, represents 195 local energy companies that deliver clean natural gas throughout the United States.

www.ans.org
> The American Nuclear Society's core purpose is to promote awareness and understanding of the application of nuclear science and technology.

www.ases.org
> The American Solar Energy Society is a leading association of solar professionals and advocates.

www.awea.org
> The American Wind Association promotes wind energy as a clean source of electricity for consumers around the world.

www.api.org
> The American Petroleum Institute is a national trade association that represents all aspects of America's oil and natural gas industry.

www.geo-energy.org
> The Geothermal Energy Association is a U.S. trade organization composed of U.S. companies who support the expanded use of geothermal energy for electrical power generation and direct-heat uses.

www.greenpeace.org
> Greenpeace is an international organization that campaigns for climate solutions that will foster prosperity without damaging the planet.

www.hydro.org
> Founded in 1983, the National Hydropower Association is the only trade association in the United States dedicated exclusively to advancing the interests of hydropower energy in North America.

www.eere.energy.gov
> The Office of Energy Efficiency and Renewable Energy (EERE) invests in clean-energy technologies that strengthen the economy, protect the environment, and reduce dependence on foreign oil.

VIDEOS

The following video and audio selections are suggested to enhance your understanding of energy topics and issues. The author has made a consistent effort to include up-to-date Web sites. However, over time, some Web sites may move or no longer be available.

Viewing some of these videos may require special software called plug-ins. Therefore, you may need to download that software to view the videos. You also may need to upgrade your player to the most current version.

- **Explanation of $E = mc^2$:** A brief explanation about the relationship between matter and energy. To learn more, go to www.youtube.com/watch?v=ejlpfOvLtI4 (02:40 minutes).
- **U.S. Energy:** To view this excellent snapshot video discussing U.S. energy use, go to www.teachersdomain.org/resource/tdc02.sci.life.eco.energyuse/ (05:00 minutes).
- **The Political Debate—Nuclear Energy versus Alternative and Carbon:** Review the Democratic debate on energy in general. For more, go to www.youtube.com/watch?v=XdMHHIO5tQM&feature=fvw (03:02 minutes).
- **Energy versus Fossil Fuels:** This video lays the foundation for fossil fuels and their impact on global warming. Go to http://videos.howstuffworks.com/hsw/6189-energy-fossil-fuels-video.htm (02:09 minutes).
- **Energy and Fossil Fuel Use:** Are fossil fuels relics of the past? An informative video that challenges the reliance on burning fossil fuels

and their place in the future of energy production. Tough questions are raised about carbon tax, the electric grid, better storage systems and battery technologies, and key enablers: www.engineering.com/Videos/VideoPlayer/tabid/4627/VideoId/1021/Energy And-Fossil-Fuel-Use.aspx (08:24 minutes).

Chapter 2

Petroleum

Petroleum has been used for thousands of years for various purposes and by various cultures. For example, early civilizations used petroleum as a medicine and as a sealant to waterproof their boats, the Chinese discovered a way to use crude oil for fuel in their lamps, and the Egyptians and Native Americans used oil to treat wounds and for other medicinal purposes.

HOW DO WE USE PETROLEUM TODAY?

Presently, petroleum dominates the world's energy scene. Oil provides 40 percent of all of the energy in industrial countries. No question, petroleum is the world's number one source of energy.

Crude oil is measured in barrels. The standard barrel of petroleum contains 42 U.S. gallons of crude oil. What products are made from a barrel of oil? According to the Department of Energy, a 42–U.S. gallon barrel of crude oil, after it is refined, produces about 19 gallons of finished motor gasoline and 10 gallons of diesel. The rest of the oil left in the barrel is used to make a variety of different products such as vitamin capsules, shampoo, bicycle tires, tennis rackets, combs, clothes, football cleats, fertilizers, pesticides, detergents, dishes, paints, food preservatives, and heart valves. The list of petroleum products is endless.

What a Barrel of Crude Oil Makes

Product	Gallons Per Barrel
Gasoline	19.4
Distillate Fuel Oil*	10.5
Kerosene-Type Jet Fuel	4.1
Coke	2.2
Residual Fuel Oil**	1.7
Liquefied Refinery Gases	1.5
Still Gas	1.8
Asphalt and Road Oil	1.4
Raw Material for Petrochemicals	1.4
Lubricants	0.4
Kerosene	0.2
Other	0.4

*Includes both home and heating oil and diesel fuel.
**Heavy oils used as fuels in industry, marine transportation, and for electric power generation.

Figures are based on average yields for U.S. refineries in 2005. One barrel contains 42 gallons of crude oil. The total volume of products made is 2.7 gallons greater than the original 42 gallons of crude oil. This represents "processing gain."

Nonrenewable energy sources such as petroleum will continue to play a major role in energy consumption at least until 2020. (*Source:* U.S. Department of Energy/Energy Information Administration)

PETROLEUM, A DOMINANT ENERGY SOURCE

According to the International Energy Outlook 2009 Report, petroleum is expected to remain the world's dominant energy source throughout 2010 to 2030. Petroleum will continue to be the primary energy source in the world's transportation sector. Approximately 68 percent of the petroleum will be used for vehicles, such as automobiles, buses, and trucks. About 25 percent of petroleum will be used by industry for the production of steel, chemicals, cement, and other products. Homes and commercial businesses also use petroleum for energy.

HISTORY OF PETROLEUM

Although the use of petroleum and natural gas is very old, the modern era of petroleum started more recently. It began in August 1859 at Oil Creek in northwestern Pennsylvania near Titusville. Edwin L. Drake, also known as Colonel Drake, used a homemade metal rig with a bit to drill down approximately 70 feet into the ground. The drill bit came up coated with oil. This was the first commercial well, and eventually it produced between

DID YOU KNOW?

Until the 1950s the United States produced nearly all the petroleum it needed. But by the end of that decade, U.S. production of petroleum could not keep up with demand. Eventually, imported petroleum constituted a major portion of petroleum used in the United States. Beginning in 1994, the United States imported more petroleum than it produced.

15 and 20 barrels of oil a day. Valuable products were made from oil produced during the late 1800s and early 1900s, the most important being kerosene used for fuel in lamps.

Then came the 20th century and the popularity of the new gas-powered engine automobile. Hundreds of thousands of automobiles and trucks were sold, and the demand for gasoline increased rapidly. Gasoline quickly became the most important product of crude oil.

Today, the exploration, drilling, and refining of petroleum continues throughout the world. Presently, Canada, Saudi Arabia, Mexico, United States, Venezuela, and Russia are the major oil-producing countries.

WHAT IS PETROLEUM?

Petroleum is a flammable, liquid fossil fuel that occurs naturally in deposits, usually underground, and is also known as crude oil. The composition of petroleum varies with locality, but it is mainly a mixture of hydrocarbons, of 5 to more than 60 carbon atoms each, with sulfur, nitrogen, and oxygen as impurities. Petroleum is liquid at Earth's surface, varying in density, and is described as heavy, average, or light. The light oils are the most valuable because they produce the most gasoline

HOW DID PETROLEUM FORM?

Oil was formed from the remains of billions of microscopic marine organisms that lived millions of years ago, long before the dinosaurs.

When these organisms died, they settled to the bottoms of lakes, rivers, streams, and even the oceans. As the years passed, the organisms were buried deeper and deeper under layers of sand and silt. The organisms finally decomposed. The heat, pressure, and bacteria in the deep layers physically

How Oil and Natural Gas is Formed

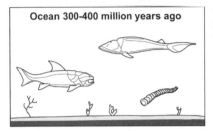

Ocean 300-400 million years ago

Tiny sea plants and animals died and were buried on the ocean floor. Over time, they were covered by layers of silt and sand.

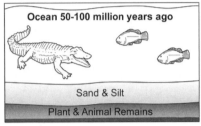

Ocean 50-100 million years ago

Over millions of years, the remains were buried deeper and deeper. The enormous heat and pressure turned them into oil and gas.

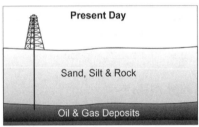

Present Day

Today, we drill down thought layers of sand, silt and rock to reach the rock formations that contain the oil and gas deposits.

(*Source:* U.S. Department of Energy/U.S. Energy Information Administration/ Illustrator: Jeff Dixon)

and chemically changed the organic remains of these organisms into a thick, dark liquid, which we now call petroleum, crude oil, or simply oil. The composition and viscosity of crude oil varies from a thin, light-colored fluid to a thick, dark-colored substance.

SEARCHING FOR OIL

Oil is not found just anywhere. Not all rock formations hold oil. Most petroleum is found in sedimentary rock basins, of which there are about 700 worldwide. Sandstone and limestone, generally folded and faulted, are common rocks where oil may be located. The oil lies in the pores of the rocks, like water in a sponge.

DID YOU KNOW?

The word "petroleum" means "rock oil" or "oil from the earth."

Much work is needed to search for these underground oil deposits. The oil search begins with a team of geologists, geophysicists, and petroleum engineers. Their job is to search, identify, and map rock formations that may contain quantities of oil and gas.

To do the work, the team uses a variety of tools, technology, and data-gathering systems to explore, locate, and map underground rock layers containing oil and gas deposits. Some of their tools for the quest for oil include satellites, global positioning systems, high-speed computer software and hardware, and 3-D and 4-D seismic imaging technologies.

DRILLING FOR OIL

Once the oil deposit is located in a particular rock layer, the oil companies begin the next stage, and that is to bring to the site a drilling rig and an oil derrick. The derrick contains all of the piping and other materials to pump out the oil from the well. The typical oil well, both on land and offshore, is about one mile deep.

Oil drillers use rotary equipment and hardened drill bits, lubricated by drilling fluids, to penetrate the earth's surface. To cut through the rock layers, various types of bits are used, depending on the hardness of the rock.

Once the bit reaches an oil or gas pool, the process of removing the oil begins. To regulate and monitor fluid flow and prevent potentially dangerous blowouts, oil drillers install a wellhead at the surface. New drilling technology has, on average, doubled the amount of oil or gas supplies developed per well since 1985.

 DID YOU KNOW?

Primitive rotary drilling rigs, such as the one Drake used, were popular in the 1880s. However, in 1901, the first modern rotary rig was used at the Spindletop oil field on a salt dome in Texas.

Horizontal and Directional Drilling

Oil and gas wells traditionally have been drilled vertically, at depths ranging from a few thousand feet to as deep as five miles. Depending on the rock layers, technology advances now allow wells to change from the standard vertical position to a completely horizontal one or even to be inverted toward the surface.

Directional and horizontal drilling enable producers to reach reservoirs that are not located directly beneath the drilling rig. This kind of drilling is particularly useful in avoiding sensitive surface and subsurface environmental features. About 90 percent of all horizontal wells have been drilled into

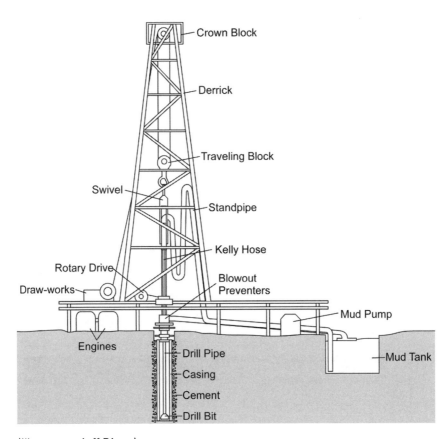

(Illustrator: Jeff Dixon)

rock layers, which account for about 30 percent of all U.S. reserves. In a given year, 40,000 wells may be drilled in the search mostly for gas reserves.

Recovery

The process by which oil or gas is removed from the well is called recovery. Pressure inside the well forces the gas or oil out of the well. However, most wells do not flow naturally, so pumps are used to force the liquids or gas out of the ground. The natural flow and the special pumps do not always remove the gas and oil in the ground.

Enhanced Recovery

A special technology called the enhanced oil recovery (EOR) technique has been developed for increasing the amount of oil that can be extracted from an oil field.

In U.S. oil fields, producers have attempted EOR techniques that offer the potential for recovering 30–60 percent or more beyond the usual amount extracted from the oil well's original site. Three major categories of EOR have been found to be commercially successful: thermal recovery, gas injection, and chemical injection.

Thermal Recovery

Thermal techniques account for more than 50 percent of all U.S. EOR production, primarily in California. Simply stated, thermal recovery involves the injection of steam into a well. The hot steam reduces the thick, heavy oil's viscosity—that is, the fluid's ability to resist flow. The steam thins out the oil, allowing it to flow through the reservoir to be recovered.

Gas Injection

Another enhanced recovery technique commonly used is called gas injection. In this technology, a gas such as carbon dioxide (CO_2) or nitrogen is injected into the reservoir. In the reservoir, the gas expands and thereby pushes additional oil up and out of the rock oil reservoir. CO_2 injection has been used successfully throughout western Texas and eastern New Mexico. Gas injection is now being evaluated as an EOR process in such states as Mississippi, Wyoming, Oklahoma, Colorado, Utah, Montana, and Alaska.

DID YOU KNOW?

In the United States, the average size of a new oil field installed on land is much smaller than years ago. These new fields do not remain productive for long. More than 50 percent of them produce fewer than 10 barrels per day.

Chemical Injection

The next type of EOR uses a detergent-like material instead of heat or gas to recover oil in a reservoir. The use of detergent-like materials reduces the surface tension of the water in the reservoir that often prevents oil droplets from moving through a reservoir. The chemical injection technology is not a major enhancement process. Today, it accounts for less than 1 percent of U.S. EOR production.

Presently, EOR reports indicate that these techniques are not implemented very often because of their relatively high costs, and in some cases, these methods are not too efficient. However, because there is a lot of U.S. oil in the ground that presently cannot be recovered, there will be more research in enhanced recovery technologies to retrieve this important resource.

CRUDE OIL REFINERIES

Once the petroleum is recovered and piped out of the well, it is transported by tankers and pipelines and shipped to refineries. The refinery is a huge complex that separates crude oil into various products, such as gasoline, kerosene, and asphalt, in a fractional distillation tower. Inside the tower, the petroleum undergoes distillation produced by steam. The oil is heated into a vapor, which rises inside the tower. Because the components in the crude oil vapor, have different boiling points, they condense back to a liquid state at different temperature levels in the tower, producing separate products such as diesel fuel, gasoline, and jet fuel. The lightest products,

VIDEO

U.S.—Extreme Drilling: This method may recover heavy, thick oil, but at what cost? To learn more, go to http://www.youtube.com/watch?v=QP2GejkLdwA&feature=related (03:03 minutes).

such as gasoline, are removed at the top of the tower. The heaviest portions are removed at the bottom. When all products are completed, they are pumped from the condensation level in the tower to special storage tanks.

THE MAIN PRODUCTS OF PETROLEUM

Gasoline

As mentioned earlier, gasoline is one of the main products of a refinery. Gasoline is a light, volatile, highly flammable mixture of hydrocarbons. These hydrocarbons are obtained in the fractional distillation of petroleum, shale oils, or coal and are used as a fuel for internal combustion engines and as a solvent. Solvents are special liquids that can dissolve other substances.

Gasoline is a complex mixture, containing hundreds of different hydrocarbons, most with 3 to 12 carbon atoms per molecule, but varying widely in structure. It is perhaps the most widely used product refined from petroleum. Gasoline is useful as an automobile fuel because it easily evaporates to a gas, which when burned releases a great deal of energy.

Presently, gasoline is rated by an octane number. The octane rating indicates how much the fuel can be compressed in an engine before it spontaneously ignites. When the wrong-rated octane gasoline is used to power a vehicle, it can cause knocking in the engine—not a good thing. At one time, until the late 1960s, to increase the octane rating, additives containing lead were widely used. Because of the health and environmental hazards, manufacturers in the 1970s began to change automobile designs and gasoline composition to exclude lead. In 1990, the Clean Air Act (CAA) forced major compositional changes in gasoline; lead additives are now banned in the United States.

Diesel Fuel

Diesel fuel is the common term for the motor vehicle fuel used in compression ignition engines. It was named for its inventor, the German engineer Rudolf Diesel, who patented his original design in 1892.

There are differences in the refining of diesel fuel and gasoline. Petroleum diesel is a "distillate" refined from crude oil. A distillate is a purified liquid produced by condensation from a vapor during distilling. Diesel fuel differs from gasoline in a number of obvious ways—it is heavier, oilier, and

A diesel-fueled bus takes passengers around the grounds at the Centennial Park in Nashville, Tennessee. (iStockphoto)

much slower to evaporate. A less obvious quality, however, is diesel fuel's efficiency. Diesel car engines get more miles per gallon than do gasoline engines. In diesel engines, there is a higher compression of the air and fuel mixture, which results in the production of more power than in their gasoline counterparts. Diesel engines are common in the United States for such vehicles as trucks, buses, boats, and tractors. In Europe, where the cost of gasoline is notably higher, diesel engines are also quite common in cars. European refineries produce more diesel fuel than gasoline, the reverse of the situation in North America.

DID YOU KNOW?

One of the fuels that Rudolf Diesel originally considered for his engine was vegetable seed oil, an idea that is now coming back as so-called biodiesel. Biodiesel can be manufactured from vegetable oils, animal fats, or recycled restaurant grease from fryolators. It is biodegradable and can reduce vehicle emissions of particulates, carbon monoxide, and hydrocarbons. For more information about biodiesel, refer to volume 4.

INTERVIEW

Green Advocate: Jason Diodati, Chemistry Teacher, Marc and Eva Stern Math and Science School, East Los Angeles, California

Go-Green Project: Powering the Cars of Tomorrow, Biodiesel

I teach chemistry at the Marc and Eva Stern Math and Science School, a public charter in the Los Angeles Alliance for College Ready Academies. The school is located on the California State Los Angeles campus and draws students predominantly from East Los Angeles. Our school was opened in 2006 with a freshman class of approximately 150 students. A majority of our students are from low-income, second-language households (Spanish), and most of them will be the first in their family to attend college. I teach chemistry to the sophomore class, and I am in charge of an after-school science club/energy club that is open to all grade levels.

Where did you grow up, and what schools did you attend?

I was born and raised in San Luis Obispo, California. I attended high school at San Luis High and transferred to the local community college (Cuesta). After two years of college, I spent the summer in Salamanca, Spain, learning Spanish, and then transferred to UCLA, where I finished after two more years with a BS in biology. After college I worked as an environmental biologist for several construction projects throughout California; then I moved to Las Vegas to build a biodiesel refinery. Later on, I realized teaching was where my heart was, so I moved back to San Luis Obispo to attend college and earned teaching credentials in biology and chemistry.

What were some of your favorite activities and subjects in high school?

Most of my favorite activities in high school did not involve class, primarily because I was addicted to sports: soccer, baseball, football, wrestling, handball, golf—the list goes on and on. I also enjoyed hanging out with friends, going to the beach, and flirting with girls. Even though I never really had any educational goals or career plans, I knew that if I worked hard, eventually I would figure things out and be able to choose a career. I didn't get excited about science until I was in college.

What interested you in seeking a career in your discipline?

I have loved to teach or to help others as far back as I can remember. I would tutor other students in high school and college. I would also help my friends during study sessions and during group work activities in class.

Describe the biodiesel project.

After seeing the enthusiasm and interest of my chemistry students when exposed to alternative energy, the English teacher (Christine Powers) and I decided to have the students work in pairs and apply for alternative energy grants. We had them read through the BP America web page and follow their rubric, and we gave them support in scientific writing and producing a budget. The students were very enthusiastic about the project, and the English teacher stated it was their highest-level writing all year.

After the summer, we found out that two of the projects had been selected for funding. One grant, for $5,000, was to install a weather station on the roof of the school and to collect and monitor wind and solar data for the possible installation of a wind turbine or solar panels. (That project was completed in addition to the one described.) The second grant was for $10,000 to purchase a car and convert it to run on waste vegetable oil. The students were very empowered and proud of their projects. They had witnessed that writing could and would get them money (large amounts of money), a teaching tool that far outweighs anything I have ever done in class! Their first question, of course, was whether they could keep it.

We started the project by purchasing a 1978 Mercedes Benz 240 Diesel car. We wanted to compare the fuel economy and emissions of diesel, biodiesel, and waste vegetable oil and create a chart to educate the school and the community. After two months of collecting data, the car (and teacher) was in a head-on collision, and the car was totaled. Thankfully, we were able to purchase a second car, a 1984 Mercedes Benz 300 Turbo Diesel, to finish the project. We had to start the data-collection process over, and we also coincided this data collection with the car conversion. The fuel economy data were charted in an Excel graph showing average miles traveled for gallons of diesel used after 10 tanks (we did this for each fuel type).

Jason Diodati, a chemistry teacher, and his students at the Stern Math Science School in California, built a biodiesel car powered solely by vegetable oil. From left to right: Andres Juarez, Daniel Ramirez, Jose Rodriguez, Jason Diodati, Angel Aleman, Cesar Gomez, Fermin Rojas, Raquel Molina, Abigail Marquez, Emily Cendejas, Janneth Cervantes, Xiomara Martinez, Ana Baez, Maricruz Gutierrez, Crystal Mendoza, Stephanie Alonzo, Rosy Palapa, and Jasmine Vidaca. (Photo by Brendan Vitt)

We purchased the Greasecar conversion kit and began working one hour a day one day a week converting the car. We also collected data on the emissions of diesel fuel. After a few months we started running the car on biodiesel that we purchased from a local gas station. We continued to collect data on fuel economy for biodiesel.

When the conversion was complete, we began collecting waste vegetable oil from local restaurants and filtering it for use in the car. The students quickly learned that getting clean oil meant less filtering, so they tried hard to find the best oil they could. We used sock filters from 100 to 10 microns prior to using it in the car.

The students then compared the fuel economy of the three fuels and analyzed the performance and emissions. We then celebrated by driving the car to Magic Mountain solely powered by vegetable oil!

How or why did you get interested in selecting this project?

Given that my background is in chemistry and biodiesel, which I learned in Las Vegas, I decided to bring this subject to my chemistry students. I started by introducing my first year of students to the concept of alternative energy and fuel. When they showed interest in the subject, I decided to team up with the English teacher to get funding for a biodiesel type project.

Explain the importance of the project as it relates to real-world issues.

Alternative energy research has become one of the most discussed topics in politics in the last few years. Both former President Bush and President Obama have expressed interest in increasing funding for research and education in this area. Many companies such as BP America have begun dedicating a large portion of their profits to this field. People have started to realize that alternative energy will be pivotal in maintaining our current demands for energy and balancing the environmental needs of the planet. Also, most governments are starting to see that alternative energy is necessary for financial survival and independence. With President Obama's new energy plan and revitalization efforts, the number of "green collar" jobs is growing exponentially, and our students need the training and desire to pursue those fields as careers.

What materials and references (Web sites, advisors, periodicals, etc.) did you use as resources for the project?

Our main resource was BP America. They provided us with the funding to complete the project. We also used Greasecar (www.greasecar.com) to purchase the kit to run the vehicle on waste vegetable oil. The rest of the information came from trial and error during the project and my background knowledge.

What advice would you give other teachers who would like to do this project?

My advice for other teachers is to make sure you have proper funding and then go for it. Our project was completed without any auto shop or auto shop teacher, without the proper facilities, without the proper tools, and often without the technical know-how of the teacher. We all worked together, and we got the job done without any errors or mistakes, and the car ran perfectly! However, without the grant from

BP America to purchase the car and other supplies, this project would never have even been considered.

Discuss some of the students' contributions to this project.

The project was completely driven by the students; I was their guide and inspector, but they were the workers and the thinkers. The best way to learn about something is to roll up your sleeves and dive in head first, which is what they did. I was immensely proud of their work and very impressed with their learning curve. Many of the students who had never held a pair of pliers were eagerly completing many of the more difficult mechanical steps of the process. Also, many of the girls were outperforming the boys in their ability to get their hands dirty and crank bolts and cut metal. Overall, the students were very excited about their creation (modification) and were the envy of the school.

How long did it take to complete the project?

Our project was completed in many steps, but overall we needed about six months to finish the entire project. Our school has no auto shop, no tools, and no place to park or store a vehicle. We also have students whose primary transportation is the bus or metro, and often it is dangerous for them to stay late after school and arrive near their homes in the evening or late afternoon. Therefore, our project consisted of one hour after school for only one day each week. We met on three different Saturdays for about four hours each time. This means that we were working very slowly, and each part of the project and car conversion had to be broken down into very small parts. The car had to be moved off campus every day, meaning the car had to be fully functional and running after every one-hour session. We had to save the major parts of the project, such as the fuel system and heating/cooling system, for the Saturdays when we had a longer period of time. Also, the students who participated in this project had never done any mechanical work, increasing the difficulty of meeting time deadlines. A school that has a shop and students with some mechanical background should be able to finish the conversion in a much shorter amount of time.

Do you have any current plans to improve on or to extend this program? If not, what are you planning next in the field of energy, conservation, or the environment?

Sadly, we did not receive funding to continue our research on alternative fuels and biodiesel. Our goal was to extend the program to build a biodiesel processor on campus and purchase a school bus. We were then going to produce our own fuel from waste vegetable oil and use the bus to transport our athletes to games and our clubs on field trips. We will have to wait for additional funding to complete this portion of the project.

The goal for the upcoming school year is to use waste vegetable oil and stirring hot plates to research various methods of producing and washing biodiesel. My goal is to make the students experts in the reaction mechanism and washing technique, which will be transferred to large-scale production if needed. Essentially, they are going to be chemists and apply their knowledge from my class to a real-world application.

Petrochemicals

The use of petrochemicals or petroleum-based products extends far beyond fuels and power for our homes, cars, and factories. The strength, durability, and flexibility of petroleum-based plastics, resins, and foams make them inexpensive, resilient, and lightweight.

Petrochemicals are compounds derived from petroleum or natural gas, nonrenewable resources that are often referred to as fossil fuels. Petrochemicals are obtained when crude oil or natural gas is refined, or separated, into gasoline, heating oil, asphalt, and other useful substances. Some petrochemicals, such as fuels, solvents, pesticides, drugs, and cosmetic preparations, are put to direct use. Most petrochemicals, however, serve as raw materials, or intermediates, in the production of synthetic substances, particularly plastics.

Ethylene, a highly reactive gas, is perhaps the most widely used petrochemical. It is used in the production of plastics, synthetic fibers, and antifreeze. Other important petrochemicals include benzene, which is used to make synthetic rubber and latex paints, and phenols, which are important chemicals used in the manufacture of perfumes, artificial flavorings, and pesticides.

Petrochemical products are used in just about every industry today, from agriculture to medicine. Unfortunately, the production and use of petrochemicals causes a variety of environmental problems. When these substances are produced, for example, a number of pollutants, including sulfur dioxide and particulates, are released into the air. Emission of sulfur dioxide is one of the main contributors to acid rain formation. Certain petrochemicals themselves, such as benzene and toluene, are also highly toxic to humans and other organisms.

Other Products

Several other products are made at refineries. These products include liquefied petroleum (LPG), kerosene, fuel oil for furnaces, aircraft fuels, and asphalt for road building.

In addition to crude oil, refineries and blending facilities add oils and liquids to produce finished products for sale to consumers. These products include such items as the flammable liquid naphtha and kerosene. Blending facilities add oxygenates (such as ethanol) and various "blending components" to produce finished motor gasoline for gas stations. Blenders also add relatively small but increasing amounts of "biodiesel" (made from

FEATURE

Alaska Pipeline

The U.S. government in 1973 approved construction of the Alaska Pipeline, an 808-mile-long pipeline that transports petroleum across the state of Alaska, providing about 10 percent of the oil used in the United States. The Alaska Pipeline, also known as the Trans-Alaska Pipeline, extends from Prudhoe Bay, near the Arctic Circle, to the Port of Valdez in southern Alaska. The 5-foot-diameter pipe carries approximately 2 million barrels of oil each day. Oil flow is maintained by 11 pump stations located along the pipeline.

To help prevent permafrost damage, more than half of the pipeline is located above ground. Where the pipeline is below ground, it is refrigerated or buried in thaw-stable, non-permafrost areas. In total, the Alaska Pipeline is buried under or crosses over more than 800 rivers and streams. Workers constructed 13 bridges along the route, including a 2,297-foot bridge that passes over the Yukon River.

The pipeline has been a subject of concern to citizens and environmentalists. One major concern was how pipeline construction and operation might impact Alaska's fragile tundra ecosystem, specifically its permafrost layer.

processed grain oils and other products) to diesel fuel and even heating fuel.

From the refinery, most petroleum products are shipped out through pipelines. There are more than 200,000 miles of pipelines in the United States used for transporting petroleum products.

U.S. PRODUCTION OF CRUDE OIL

Oil-Producing States

Approximately 25 percent of all crude oil produced in the United States comes from offshore drilling rigs in the Gulf of Mexico. Texas, Alaska, California, and Louisiana are responsible for 52 percent of total U.S. crude oil production. Other states that produce crude oil are Oklahoma, North Dakota, South Dakota, Montana, Florida, Mississippi, Alabama, and Nebraska.

CRUDE OIL IMPORTS TO THE UNITED STATES

Because the United States consumes about twice as much crude oil as it produces, it must import supplies from other countries. The U.S. imports crude oil from over 60 countries, and in 2000, approximately 70 percent of net imports of petroleum were from five countries: Canada, Saudi Arabia,

FEATURE

Long Beach, California

Believe it or not, the petroleum industry also drills and produces crude oil and gas in the midst of some of the nation's largest cities. In recent decades, the industry has successfully produced crude oil in urban environments where operations are frequently visible for all to see.

One urban area where the production of crude oil takes place is Long Beach, California. The oil company runs a 43,000-barrel-per-day operation at the East Wilmington unit, located in the city of Long Beach's scenic harbor. The East Wilmington unit is part of the giant Wilmington oil field, one of the nation's largest.

Production at East Wilmington occurs on four human-made islands built on 640,000 tons of boulders and 3.2 million cubic yards of sand dredged from the harbor and concealed by palm trees, flowers, concrete sculptures, waterfalls, and colorful nighttime lighting. These islands represent the centerpiece of a solution between the industry and the city of Long Beach to tap the harbor's resources without harming its natural beauty.

To shelter operations from public view, drilling rigs are covered by structures. They are built to resemble high-rise buildings, and wellheads and other support facilities are located below ground.

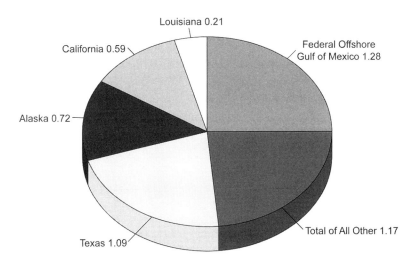

U.S. Crude Oil Production 2007 by Major Producing States and Federal Gulf of Mexico (Million Barrels per Day)

- Louisiana 0.21
- California 0.59
- Federal Offshore Gulf of Mexico 1.28
- Alaska 0.72
- Texas 1.09
- Total of All Other 1.17

Texas and Alaska are the leading oil-producing states in the United States. (*Source:* U.S. Department of Energy/Energy Information Administration, *Petroleum Supply Annual 2007*)

Global Issues—Peak Oil: The world faces increasing demand for and a shrinking supply of a finite resource. To learn more about the ultimate fuel supply crisis, go to http://www.youtube.com/watch?v=DMQd5nGEkr4 (10:00 minutes).

Mexico, Venezuela, and Nigeria. Imports from all OPEC countries made up 49 percent of total U.S. crude oil imports.

Major Oil-Producing Countries

As of 2007, 10 countries produced 60 percent of the total world production of oil. The top five, which produced 42 percent of the world total, and their share of total world production were as follows:

- Russia, 13 percent
- Saudi Arabia, 12 percent
- United States, 7 percent
- Iran, 5.4 percent
- China, 5.1 percent

OPEC

Several countries that make up an organization called Organization of Petroleum Exporting Countries (OPEC) produce much of the world's oil. OPEC includes Algeria, Angola, Ecuador, Indonesia, Iran, Iraq, Kuwait, Libya, Nigeria, Qatar, Saudi Arabia, the United Arab Emirates, and Venezuela. In 2007 OPEC produced more than 40 percent of the world's oil.

Oil reserves are increasingly concentrated in OPEC countries and by national oil companies. These 13 countries control nearly 80 percent of the world's oil reserves. Most oil analysts agree that OPEC members are major players in the world oil market and can set prices as world oil demand rises.

DID YOU KNOW?

The United States has no national oil company. The largest three U.S.-based international oil companies are ExxonMobil, Chevron, and ConocoPhillips.

OIL SHALES AND OIL SANDS

Two kinds of materials can be converted into a petroleum-like liquid: oil shales and oil sands. Oil shales are rock-like in appearance and oil sands look like the soft asphalt that is used for driveways and roads. Both contain organic chemicals that can be refined for fuels.

Oil Shales

The term "oil shale" generally refers to any sedimentary rock that contains solid bituminous materials called kerogen. The kerogen is a mixture of organic chemical compounds that make up a portion of the organic matter in sedimentary rocks.

Kerogen can be converted into petroleum-like liquids when the rock is heated in the chemical process. Oil shales can yield 25 gallons of oil per ton of rock when heated. Kerogen can be produced as a superior-quality jet fuel, diesel fuel, and other products as well.

How Was Oil Shale Formed?

The Oil Shale and Tar Sands Programmatic Environment Impact Statement (PEIS) describes how oil shale was formed: "Oil shale was formed millions of years ago by deposition of silt and organic debris on lakebeds

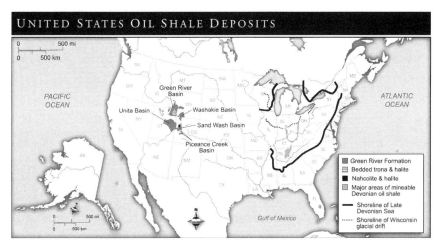

Shale deposits have been discovered in several areas of the United States. (*Source*: U.S. Geological Survey, Geology and Resources of Some World Oil-Shale Deposits, Scientific Investigations Report 2005-5294)

and sea bottoms. Over long periods of time, heat and pressure transformed the materials into oil shale in a process similar to the process that forms oil. However, the heat and pressure were not as great. Oil shale generally contains enough oil that it will burn without any additional processing, and it is known as 'the rock that burns.'"

Mining of Oil Shale

As described in the PEIS, "oil shale can be mined and processed to generate oil similar to oil pumped from conventional oil wells. However, extracting oil from oil shale is more complex than conventional oil recovery and currently is more expensive. The oil substances in oil shale are solid and cannot be pumped directly out of the ground. The oil shale must first be mined and then heated to a high temperature." This process produces a liquid that goes to a refinery plant where it is processed as a synthetic crude oil.

In 1912 the U.S. government established the Naval Petroleum and Oil Shale Reserves. This office has overseen the U.S. strategic interests in oil shale. Since that time, some commercial attempts have been made to produce oil from oil shale. However, these attempts have failed primarily because the cost of petroleum was lower.

Meaningful federal oil shale policy initiatives have not been undertaken since the 1980s. In that time technology has advanced, and global economic, political, and market conditions have also changed. In the United States, the richest oil shale deposits are located in Colorado, southeastern Utah, and southern Wyoming. The nation's total oil shale resources could exceed 6 trillion barrels of oil. However, production of the oil shales may not be economically feasible.

Oil Sands

Oil sands are layers of sticky, tar-like bitumen that are mixed with sand, clay, and water. Bitumen is a black, oily, liquid material that is a byproduct of decomposed organic materials. Bitumen is also known as asphalt or tar. Bitumen has been mixed with other materials throughout history and has been used as a sealant, adhesive, and building mortar and as a decorative application on pottery.

Deposits of oil sands occur in several areas of the world, including the United States, Russia, and the Middle East. The lion's share of these deposits, however, can be found in only two countries: Canada and Venezuela. Astonishingly, the oil sands reserves in these two countries are roughly equal to the world's remaining total reserves of conventional crude oil. The oil sands in Alberta, Canada, cover an area the size of North Carolina.

Mining Oil Sands

When oil sands are mined, approximately one hundred feet of topsoil must be removed to reach the oil sand deposits. Then the production of extracting the oil sands starts at the bottom of the mine. Here a three-story giant caterpillar machine with a shovel scoops up the sand and dumps it into huge trucks. Each truck, carrying approximately 400 tons of the oil sand, transports the material to an extraction plant. In the plant, the bitumen in the oil sand is washed out and separated from the sand in a bath of hot water. Leaves and other debris in the wash are removed. The bitumen material is sent to a refinery plant that converts it to synthetic crude oil. The leftover water and debris are piped to a special pond where the water is treated and cleaned to be reused in the mines.

Despite the complexity of this extraction process, oil sands have created a so-called black gold rush in the Canadian province of Alberta. The Canadian government and oil companies have been quite successful in drawing oil from Alberta's sands, producing about 1.1 million barrels daily. Having committed $80 billion more to intrastructural development, government and industrial investors hope the sands will be yielding 5 million barrels per day by 2015.

U.S. IMPORTS OTHER THAN REFINED CRUDE OIL

In addition to crude oil, the United States also imports 2 million barrels per day of refined products. The refined products are special blends of petroleum fuels such as fuel ethanol. Five countries accounted for 71 percent of the U.S. imports of refined petroleum products in 2007: Canada, Virgin Islands, Russia, Algeria, and Venezuela.

 DID YOU KNOW?

At room temperature, bitumen is similar to the texture of molasses. Below 50 degrees, bitumen is a very hard rubber-like material.

THE STRATEGIC PETROLEUM RESERVE

The United States created the Strategic Petroleum Reserve in the late 1970s with the goal of protecting the country from disruptions in the oil supply. The reserve held 493 million barrels by 1985, enough oil to replace about 115 days of net petroleum imports. Although the reserve held 541 million barrels in 2000, that amount now would replace only 53 days worth of imports in large part because of increased energy consumption.

ENVIRONMENTAL ISSUES

When large volumes of petroleum products are burned, greenhouse gases and other byproducts are released into the air. These gases can cause serious

Crude oil pipes at the Strategic Petroleum Reserve Bryan Mound site near Freeport, Texas. (U.S. Department of Energy)

 DID YOU KNOW?

Of every 10 barrels of petroleum consumed in the United States in 2000, more than 4 barrels were consumed in the form of motor gasoline. The transportation sector alone accounted for two-thirds of all petroleum used in the United States in 2000.

FEATURE

Biofuels

Biofuels are used as an alternative to fossil fuels and include biogas, biodiesel, and methane. A biofuel is a solid, liquid, or gaseous fuel derived from biomass. About 5 percent of the energy consumed in the United States is provided by biofuels. Most of the biofuels are produced from wood waste from logging operations, but they can also be produced from corn and sugar crops. In France, Italy, and Germany, biodiesel fuels are produced from domestic oilseeds and cottonseeds. Biofuels are cleaner than fossil fuels because they release fewer emissions. To learn more about biofuels, see volume 4 in this series.

environmental problems such as air pollution, smog, acid rain, and an enhanced greenhouse effect. Scientists generally believe that the combustion of fossil fuels and other human activities are the primary reasons for the increased concentration of carbon dioxide in the atmosphere.

Petroleum products burned to run vehicles, heat buildings, and provide power for factories are responsible for about 80 percent of the world's CO_2 emissions, about 25 percent of U.S. methane emissions, and about 20 percent of global nitrous oxide emissions.

Processing petroleum and using its products also create many other air pollutants, including airborne particulate matter. In addition, oil spilled from tankers and offshore wells has damaged ocean and coastline environments. The environmentally disruptive effects of oil wells also have sometimes led to strong opposition to new drilling, as in wilderness areas of northern Alaska and in the Arctic National Wildlife Refuge (ANWR).

Oil spill pollution is a particular problem in the world's oceans, where it can have devastating effects on wildlife and ecosystems. Oil is toxic and directly kills small animals, such as fishes, birds, shrimp, crabs, and other shellfish.

One major oil spill took place on Tuesday, April 20, 2010. An offshore oil drilling platform, Deepwater Horizon, exploded in the Gulf of Mexico near Louisiana. It was the largest accidental marine oil spill in the history of the petroleum industry. The spill has caused extensive damage to marine and wildlife habitats, and the Gulf's fishing and tourism businesses have also suffered. The Deepwater Horizon well was capped on July 15, 2010, and is no longer discharging oil into the Gulf of Mexico, according to the Florida Environmental Protection Agency (EPA).

Today, laws help protect against oil pollution in the oceans. For example, the federal Oil Pollution Act of 1990 increases the legal liability of oil tanker owners by requiring them to adhere to strict regulations regarding oil transport. The petroleum companies have also made gains in reducing emissions at their refineries.

HOW MUCH OIL IS LEFT?

Many in the petroleum business have estimated that about 1.3 trillion barrels of oil still exist presently in known oil reserves. If that number of barrels is accurate, the data show that today's reserves will last a little more than 40 years. Other experts believe that oil production has already reached its peak as of 2009 and that reserves will last less than the 20 years or so.

However, some experts believe that new and advanced technologies can be applied to extract more oil from hard-to-find places like underground reservoirs. One new idea is to use a third stage of extraction to pump up more oil from known and existing oil reservoirs. If these technologies can be developed, there is a possibility that oil could last at least another century. Following are the current stages and proposed third stage of oil extraction.

First stage. Presently, according to experts, about 10–15 percent of all oil in a reservoir is ejected to the surface after drilling. The internal pressure in the well pushes the oil out. This is called the first stage of oil recovery.

Second stage. Once the oil can no longer be pumped out by internal pressure in the well, oil companies use a second stage of recovery. In this stage, water or natural gas is injected into the ground, providing enough pressure to force more oil out of the well. Between the first stage and this stage, about 20–40 percent of the original oil is extracted. The remaining oil (about 60%) is trapped in smaller pockets or is too thick to flow toward

 DID YOU KNOW?

Most of the oil and gas producers are independent oil and gas small businesses, typically employing, on average, 10 full-time and 3 part-time employees. These companies drill 85 percent of the nation's wells and produce 65 percent of the natural gas and nearly 40 percent of the oil consumed by Americans.

the reservoirs or into the wells on its own. So presently all of this oil is unrecoverable.

Third stage. In order to extract the remaining 60 percent in the original well, new technologies have to be adopted. Now oil companies are experimenting with a new stage to recover the remaining oil. This technology is still very new and expensive. More experimentation and studies need to be conducted. But here, briefly, are some of the proposed methods for a third stage of extracting oil:

- **Incendiary.** This method would include burning part of the reservoir to produce heat to thin out the oil so that it could flow easily to the surface. The heated air would also produce carbon dioxide, a gas that would add pressure to force the oil to the surface.
- **Chemical.** Special chemicals called surfactants would be injected into the oil reservoir. Surfactants are materials that lower the surface tension of a liquid, allowing the liquid to flow easily and quickly. These kinds of chemicals would assist the oil in breaking away from the rocks to flow better. This is similar to the way we use soap to wash our dishes.
- **Biological.** In this method, bacteria would be injected into the reservoir, where the microorganisms would grow between the oil and the rock. In time, the oil would then be released from the rocks to be extracted.

THE FUTURE OF PETROLEUM

As stated earlier, petroleum will continue to dominate fuel in the energy marketplace. The U.S. Census Bureau's world population report states that the population is expected to increase steadily over the first half of the 21st century. More people means more demand for fuel, energy, plastics, and food—all highly dependent on oil. In the 10 years from 2002 to 2012, the world population is expected to increase from 6.23 billion to 6.96 billion, an extra 12 percent to be fed. Along with population, another factor is the increasing use of oil in developing countries—countries that, up to now, had been contributing little to consumption. Therefore, the problems of oil consumption in the future revolve around two factors: population and the increasing use by developing countries that want to obtain economic progress for their people.

SPECIAL INTERVIEW: CAREERS IN ENERGY

Green Advocate: Keats Moeller, ConocoPhillips Company, Houston Texas

In this interview, Keats Moeller discusses her career in the energy field. If you wish to learn more about energy-related careers, refer to the special section in the appendix "Opportunities in Renewable and Nonrenewable Energy Careers."

The publisher and the author wish to thank National Energy Education Development (NEED) for permission to use the following Keats Moeller interview from their newsletter *Career Currents*. To learn more about NEED, visit their web site at www.need.org

Career Chat: Human Resources

Keats Moeller is a Senior Advisor of Recruiting & Staffing for the ConocoPhillips Company in Houston, TX. Keats holds several degrees: a BBA in Marketing and a MS in Management from Texas A&M in College Station, TX, and an MBA in Finance/Marketing from Southern Methodist University in Dallas, TX.

Describe what you do.

I work in the university-recruiting group at ConocoPhillips. I have an exciting job working with, and identifying, the next generation of top talent for ConocoPhillips. I have the opportunity to meet and work with college students from across the nation and around the world.

ConocoPhillips has an amazing internship program for college students from a variety of disciplines such as accounting and finance, marketing, communications, human resources, engineering and geo-sciences. While college students are at ConocoPhillips for their summer internships, the focus is on a meaningful assignment. However, ConocoPhillips also pairs each student with a mentor and provides learning opportunities, to introduce students to other new hires and to experts in our company. I coordinate speakers for these events on topics such as the energy landscape, sustainable development and learning about different aspects of our company such as exploration, production, and refining, and the importance of communicating our energy policy. In addition, I coordinate community service activities so that our interns are able to participate in a hands-on way.

Describe your typical day of work.

When I first get into my office in the morning, I review my calendar for the day, and then catch up on e-mails and voicemail. From there, my days vary and may include preparing for our university recruiting activities, preparing for our summer interns, or even working with faculty and staff at universities.

Did any special course work or training help you gain your current position?

There are many ways I prepared for my current job. Throughout school, I took a variety of courses that gave me a broad exposure to business and science. In addition,

I got involved in a number of student organizations, which was a great way to build my ability to work in teams and to hold leadership roles. One of the most valuable experiences I had while in school was working in internships in my field of study. Internships gave me the opportunity to get a first-hand look at companies and to apply what I had learned. The great thing was that I was able to make a real contribution to a company—and then when I went back to school, my coursework made even more sense! Once I was in industry, I have had the opportunity to continue learning through each of the jobs that I have held.

Please share some of the opportunities you've had.

One of the most amazing opportunities has been all of the people that I have met from college students and new employees to our experienced employees. I've even had the opportunity to travel with the CEO, Vice President and the Controller of our company.

What challenges do you face working in human resources?

One challenge of my profession is identifying and preparing for the workforce of tomorrow. The energy industry is an exciting and interesting place that is always changing. Students have a number of choices of where to begin their career. My challenge is in sharing the opportunities available in our company and industry! Visit www.conocophillips.com/car eers/UnivRecruit/index.htm for information on university recruiting.

What's the most rewarding part of your job?

The most rewarding part of my job is working with students. I truly believe in the opportunities that we have to offer at ConocoPhillips. It is exciting to be able to introduce and share our company with our newest employees.

What's the most surprising part of your job?

One of the most surprising aspects of my job has been to see the number of employees we have who have been at our company for over 20 years—and how engaged these employees are and how willing they are to share their experiences with new and potential employees.

Would you follow the same career path again?

The energy industry is an exciting place to work. One of the most rewarding aspects is knowing that I am involved in a company that is making a difference in the world we will live in. I feel very fortunate to work with the phenomenal talent that ConocoPhillips attracts.

Careers in energy are interesting and dynamic and I would definitely encourage young people to consider these careers! My advice would be to focus on doing well in school and to get involved in organizations as a way to learn more about teamwork, leadership, and an area of interest. Once you are in a company, your technical skills are important, but so is your ability to interact with others and to build relationships.

BOOKS AND OTHER READING MATERIALS

Adelman, Morris A. *The Economics of Petroleum Supply*. Cambridge, MA: MIT Press, 1993.

Magueri, Leonardo. *The Age of Oil: The Mythology, History, and Future of the World's Most Controversial Resource*. Westport, CT: Praeger, 2006.

Nakaya, Andrea, ed. *Oil: Opposing Viewpoints*. San Diego, CA: Greenhouse Press, 2006.

Raymond, Martin, and William Leffler. *Oil: Beginner's Guide*. Oxford, UK: One World, 2008.

Smil, Vaclav. *Oil: Beginner's Guide*. Oxford, UK: One World, 2008.

SOMETHING TO DO

1. On a world map locate those geographic areas—land and water—where crude oil has been discovered and extracted. Research and report on those areas that have experienced serious environmental setbacks from oil drilling and/or its delivery systems (supertankers, pipelines, off-shore platforms).

2. Explore and report on the economic and political impact of having a small number of countries controlling a natural resource, such as petroleum, that is needed on a regional or global basis.

3. In the late 1950s, construction began on America's vast interstate highway system. Explore the impact of its completion on freight distribution, population shifts, mass transit systems, nonrenewable fuel consumption, air quality, and global warming. Use the following Web sites for references and research: www.eia.doe.gov and www.epa.gov.

WEB SITES

The following Web sites, though not inclusive, include government and nongovernmental organizations.

www.api.org
   The American Petroleum Institute represents producers, refiners, marketers, and transporters of oil, natural gas, and products. Its Web site includes extensive information on topics such as the history of oil, materials for teachers, petroleum products at home, and petroleum museums.

www.ogj.com/index.html
   The *Oil and Gas Journal* focuses on new developments in the oil and natural gas businesses.

www.usgs.gov/
> The U.S. Geological Survey is a science organization that provides impartial information on the health of our ecosystems and environment, the natural hazards that threaten us, the natural resources we rely on, the impacts of climate and land-use change, and the core science systems that help us provide timely, relevant, and useable information. The Geological Survey Energy Resources Program provides energy publications and data.

www.usachoice.net/drakewell
> The Pennsylvania Historical and Museum Commission Drake Well Museum is located on the site where Edwin L. Drake drilled the first commercially successful oil well in 1859. The museum documents the birthplace of the petroleum industry with exhibits, operating field equipment, an extensive photograph collection, and a research library.

www.capitalchevron.com
> Chevron is one of the world's largest integrated energy companies. Headquartered in San Ramon, California, Chevron conducts business worldwide. Check out the Exploration Zone for an educational primer about oil production and refining for students and teachers.

www.bp.com/
> British Petroleum (BP) is one of the world's largest energy companies, providing its customers with fuel for transportation, energy for heat and light, retail services, and petrochemical products for everyday items. BP offers educational materials on energy in six languages for teachers and students worldwide as part of its Science across the World Web site.

www.conocophillips.com
> ConocoPhillips offers several products and services that provide businesses with high-quality fuels, lubricants, chemicals, specialty products, and other solutions. Conoco's Web site features short on-line movies about some of the industry's newest and most innovative technologies.

http://www.stategeologists.org
> The Web site of the Association of American State Geologists includes an Earth Science Education Source Book for teachers and students, a compendium of earth science education materials, and services available from the 50 State Geological Surveys in the United States, Puerto Rico, and the Association of American State Geologists.

www.agiweb.org

> The American Geologic Institute is a nonprofit federation of 32 geoscientific and professional associations that represent more than 100,000 geologists, geophysicists, and other earth scientists. The Institute maintains a Clearinghouse for Earth Science Education with an electronic database for elementary school teachers and secondary school science teachers.

www.agu.org

> The American Geophysical Union is an international scientific society of more than 35,000 researchers, teachers, and science administrators in more than 115 countries, over 30 percent of whom are outside the United States. The society is dedicated to advancing the understanding of Earth and its environment and making results available to the public.

www.nef1.org

> The National Energy Foundation is a nonprofit provider of educational materials and programs related to energy, natural resources, and the environment. It is supported by businesses, government agencies, professional associations, and the education community.

http://science.howstuffworks.com/energy-channel.htm

> See the HowStuffWorks Web site library for numerous energy production videos and articles.

VIDEOS

The following video and audio selections are suggested to enhance your understanding of energy topics and issues. The author has made a consistent effort to include up-to-date Web sites. However, over time, some Web sites may move or no longer be available.

Viewing some of these videos may require special software called plug-ins. Therefore, you may need to download that software to view the videos. You may need to upgrade your player to the most current version.

> **Petroleum (Oil)—Where It Is Located:** With oil (and gas) getting harder to find, this video comprehensively discusses and demonstrates how reserves were formed and where they might be found. To learn more about the complex riddle of discovering new reserves and the technologies and disciplines involved, go to http://www.metacafe.com/watch/yt-_hwzJUDWIQQ/new_oil_and_gas_exploration/ (09:54 minutes).

Europe—Petroleum Issues: Is power without fuel Europe's answer to the developing petroleum crisis? In terms of advertising, this is a powerful and compelling video comparing petroleum with wind power: http://www.metacafe.com/watch/2687883/no_fuel_viable_energy/ (04:03 minutes).

Global Issues—Peak Oil: The world faces increasing demand for and a shrinking supply of a finite resource. To learn more about the ultimate fuel supply crisis, go to http://www.youtube.com/watch?v=DMQd5nGEkr4 (10:00 minutes).

U.S.—Extreme Drilling: This process may recover heavy, thick oil, but at what cost? To learn more, go to http://www.youtube.com/watch?v=QP2GejkLdwA&feature=related (03:03 minutes).

U.S.—Chevron Deep-Water Division: View this Reuters report on drilling deeper than ever for oil in the Gulf of Mexico, below 10,000 ft: http://www.youtube.com/watch?v=OnYELEO1UVM&feature=related (02:05 minutes).

Chapter 3

Natural Gas

The Chinese were using natural gas as an energy source more than 3,000 years ago. They ignited the natural gas to produce heat to evaporate pools of brine water to make salt. To extract the natural gas, the Chinese dug wells that extended about 1,500 feet deep. To transport the gas to the surface, the Chinese used pipes made of bamboo.

Today, natural gas is used extensively in residential homes, commercial businesses, and industrial plants in the United States. In fact, natural gas is the dominant energy used for home heating. More than 66 million homes in the United States use natural gas. The use of natural gas is also rapidly increasing in electric power generation and cooling.

Worldwide, natural gas remains a key energy source for the industrial sector and for electricity generation. This industrial sector is often divided into such industries as airplane manufacturing; steel production; and automobile, textile, consumer product, and electronic product manufacturing.

Many of these industries consume large quantities of energy and require factories and machinery to convert the raw materials into goods and products. They also produce waste materials and waste heat that may pose environmental problems or cause pollution. The industrial sector is the world's largest consumer of natural gas and is expected to account for 43 percent of projected gas use in 2030.

In the United States, natural gas ranks number three in energy use, right after petroleum and coal. About 22 percent of the energy we use in the United States comes from natural gas.

Industry is the biggest consumer of natural gas, using it mainly as a heat source to manufacture goods. Industry also uses natural gas as an ingredient in fertilizer, photographic film, ink, glue, paint, plastics, laundry detergent, and insect repellents. Synthetic rubber and human-made fibers such as nylon also could not be made without the chemicals derived from natural gas.

Residences—people's homes—are the second-biggest users of natural gas. Six in 10 homes use natural gas for heating. Many homes also use gas water heaters, stoves, and clothes dryers. Natural gas is used so often in homes because it is clean-burning.

Like residential use, commercial use of natural gas is mostly for indoor space heating of stores, office buildings, schools, churches, and hospitals.

Natural gas is also used to make electricity—it is the third-largest producer of electricity after coal and uranium. Many people in the energy industry believe natural gas will play a bigger role in electricity production as the demand for electricity increases in the future.

Natural gas power plants are cleaner than coal plants and can be brought online very quickly. Natural gas plants produce electricity about 20 percent more efficiently than new coal plants, and they produce it with fewer emissions. Today, natural gas generates 15 percent of the electricity in the United States.

To a lesser degree, natural gas is becoming popular as a transportation fuel. Natural gas can be used in any vehicle with a regular internal combustion engine, although the vehicle must be outfitted with a special carburetor and fuel tank.

WORLD CONSUMPTION OF NATURAL GAS

Worldwide consumption of natural gas is projected to increase by nearly 64 percent between 2004 and 2030. Among the end-use sectors, the industrial sector remains the largest consumer of natural gas worldwide, accounting for 42 percent of the total expected increase in demand for natural gas between 2004 and 2030. Natural gas also is expected to remain an important energy source in the electric power sector, particularly for new generating capacity.

Natural gas power plant near Ventura, California. The burning of natural gas at the power plant produces nitrogen oxides and carbon dioxide, but in lower quantities than burning coal or oil. (Georg Henrik Lehnerer/Dreamstime.com)

Electricity generation will account for 35 percent of the world's natural gas consumption in 2030. Why the increase in natural gas usage? Natural gas, although a fossil fuel, is an attractive choice for new power plants because of its high fuel-efficiency rating and lower carbon dioxide emissions than other fossil fuels. Also, there have been improvements in pipelines and new technologies for locating and drilling new wells.

HISTORY OF NATURAL GAS

In the United States, natural gas was first used to light the town of Fredonia, New York, in 1821. However, the fuel's use remained localized over the next century because long-distance transportation of gases was difficult. For most of the 1800s, natural gas was used almost exclusively as a fuel for lamps. Because there were no pipelines to bring gas into individual homes, most of the gas went to light city streets.

After the 1890s, however, many cities began converting their street lamps to electricity. Gas producers began looking for new markets for their product. It took the construction of pipelines to bring natural gas to new markets. Throughout the 1950s and 1960s, thousands of miles of pipeline were constructed throughout the United States. Today, the U.S. pipeline network, laid end-to-end, would stretch to the moon and back twice.

HOW IS NATURAL GAS FORMED?

Natural gas was formed in Earth's crust over millions of years by the chemical and physical alteration of organic matter. Oil and natural gas were created from organisms that lived in the water and were buried under ocean or river sediments. Long after the great prehistoric seas and rivers vanished, heat, pressure, and bacteria combined to compress and "cook" the organic material under layers of silt.

In most areas, a thick liquid called oil formed first, but in deeper, hot regions underground, the cooking process continued until natural gas was formed. Over time, some of this oil and natural gas began working its way upward through the earth's crust until they ran into rock formations called "caprocks" that are dense enough to prevent them from seeping to the surface. It is from under these caprocks that most oil and natural gas is recovered today.

THE CONTENTS OF NATURAL GAS

The hydrocarbons in natural gas have one to four atoms of carbon each; at Earth's surface, these compounds exist as gases. Natural gas is a mixture of flammable gases, including methane, ethane, propane, and butane. Because natural gas has no smell of its own, a substance is added to natural gas to produce an odor so that gas leaks can be detected. The mixture is usually composed of 70–80 percent methane.

Methane is an odorless, gaseous hydrocarbon formed by the thermal decomposition or anaerobic decomposition of organic matter. It is the simplest, lightest, most abundant hydrocarbon. It occurs naturally as the chief component of natural gas, in association with coal beds, and as the marsh gas released by the anaerobic bacterial decomposition of vegetable matter buried in wetland soils.

Methane is combustible and can form explosive mixtures with air at concentrations between 5 and 14 percent; explosions of such mixtures have been the cause of many coal mine disasters. As a component of natural gas, methane is used for fuel and also in making solvents and certain Freons, a trade name for chlorofluorocarbons (CFCs). Methane is a greenhouse gas whose concentration in the atmosphere has increased sharply as a result of human activities; in fact, the increase of global methane has essentially kept pace with increases in world population. Large quantities of methane are believed to be released by rice paddies, where vegetation rots in the waterlogged soils. The mining of coal and production of natural gas produce lesser amounts of methane, previously trapped with these deposits.

Other hydrocarbon constituents include ethane and propane, which are used as nonrenewable fuels. However, the composition of natural gas varies according to locality; minor components may include carbon dioxide, nitrogen, hydrogen, carbon monoxide, and helium. It is the cleanest-burning of the fossil fuels, yielding little more than carbon monoxide, carbon dioxide, and water as combustion products. Although some natural gases can be used directly from the well without treatment, most must be processed first to remove undesirable constituents, such as hydrogen sulfide and other sulfur compounds.

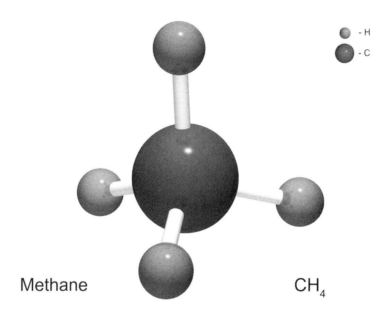

The natural gas that is delivered to many homes and businesses is almost pure methane. Methane is a molecule made up of one carbon atom and four hydrogen atoms and is referred to as CH_4. (Vasilyev/Dreamstime.com)

VIDEO

General (Global Uses): Do you know the fundamental difference between propane and butane? To discover how one carbon atom makes the difference, go to http://www.ehow.com/video_4756915_what-difference-between-propane-butane.html (01:29 minute video).

LOCATING NATURAL GAS DEPOSITS

Natural gas is often found in solution with petroleum in sand, sandstone, and limestone deposits. In order for gas to accumulate, it must be trapped. The underground gas reservoir must be sealed at the top by an impermeable stratum or cap rock, such as clay or salt. The entire cover structure must be shaped in such a way as to prevent gas from leaking to the surface. Gas accumulations are mostly encountered in the deeper parts of sedimentary basins. On the Gulf Coast of the United States, for example, more than half of the deposits discovered at depths greater than 10,000 feet are gas fields. Among the largest accumulations of natural gas are those in Siberia, the Texas Panhandle in the United States, an area in the Netherlands, and Hassi R'Mel in Algeria.

DRILLING FOR NATURAL GAS

One of the drilling methods for natural gas is known as rotary drilling and consists of a sharp, rotating metal bit used to drill through Earth's crust. This type of drilling is used primarily for deeper wells that may be under very high pressure.

According to the Natural Gas Organization, drilling for natural gas offshore, in some instances hundreds of miles away from the nearest land area, poses a number of different challenges in comparison with drilling onshore. The actual drilling mechanism used to drill into the sea floor is

DID YOU KNOW?

The digestive systems of cattle and other grazing livestock are another major source of methane. It is estimated that the large global cattle population and increasing land areas covered by rice paddies now account for almost 50 percent of the global release of methane; another 20 percent is produced by the burning of wood and other vegetation.

much the same as can be found on an onshore rig. But, with drilling at sea, the sea floor can sometimes be thousands of feet below sea level, as mentioned earlier. Therefore, although with onshore drilling the solid ground provides a platform from which to drill, out in the ocean a different drilling platform must be constructed. Since 1947, offshore production, particularly in the Gulf of Mexico, has resulted in the discovery and delivery of a great number of large natural gas deposits.

New Technologies for Drilling for Natural Gas

According to the Department of Energy, although more than 70 percent of the natural gas produced in the United States already comes from wells at 5,000 feet or deeper, only 7 percent comes from formations below 15,000 feet. Yet, at these deeper depths, an estimated 125 trillion cubic feet of natural gas may be available.

Drilling at these depths is expensive and requires new drilling technologies. For wells deeper than 15,000 feet, as much as 50 percent of drilling costs can come from penetrating the last 10 percent of a well's depth. The rock is typically hot, hard, abrasive, and under extreme pressure. In deeper wells, it is not uncommon for the drill bit to slow to only two to four feet per hour, at operating costs of tens of thousands of dollars a day for a land rig. For deep offshore drilling, the costs can exceed millions of dollars a day. And it is exceedingly difficult to control the precise trajectory of a well when the drill bit is nearly three miles below the surface.

The U.S. Department of Energy's Office of Fossil Energy has plans to help develop the high-tech drilling tools that the gas industry needs to use for deep drilling operations. The major goal of the plan is to develop a "smart" drilling system tough enough to withstand the extreme temperatures, pressures, and corrosive conditions of deep reservoirs, yet economical enough to make the gas affordable to produce.

DELIVERY OF NATURAL GAS

Methods of pipeline transportation were developed in the 1920s, and between World War II and the 1980s there was a period of tremendous residential and commercial expansion that relied increasingly on the use of pipeline transportation of gas. North American gas pipelines now extend from Texas and Louisiana to the northeast coast, and from the Alberta gas fields to the Atlantic seaboard.

The Natural Gas Industry

Production
- Oil and Gas Well
- Gas Well
- Separation
  - Water
  - Oil
  - Vented and Flared

Transmission
- Gas Processing Plant
  - Products Removed
  - Nonhydrocarbon Gases Removed
  - Returned to Field
  - Vented and Flared
- Compressor Stations
- Main Line Sales
- Odorant
- Underground Storage Reservoir

Distribution
- LNG Storage
- Natural Gas Company
- Consumers: Commercial, Transportation, Residential, Electrical, Industry

(Illustrator: Jeff Dixon)

Natural gas is delivered to customers through a safe, sound, 2.2-million-mile underground pipeline system that includes 1.9 million miles of local utility distribution pipes (1.1 million miles of utility mains, plus 800,000 miles of utility service lines) and 300,000 miles of transmission lines.

HOW IS NATURAL GAS MEASURED?

We measure and sell natural gas in cubic feet (volume) or in British thermal units (Btus; heat content). Heat from all energy sources can be measured and converted back and forth between Btus and metric units.

One Btu is the heat required to raise the temperature of one pound of water one degree Fahrenheit. Ten burning kitchen matches release 10 Btu. A candy bar has about 1,000 Btu. One cubic foot of natural gas has about 1,031 Btu. A box 10 feet deep, 10 feet long, and 10 feet wide would hold one thousand cubic feet of natural gas.

GLOBAL NATURAL GAS RESERVES

On a worldwide scale, the deposit with the greatest accumulation of natural gas reserves is a land area between the countries of Qatar and Iran. Russia is home to the second-largest deposit of natural gas reserves, which are located in various sections of this large country. Russia is also the leader in global production of natural gas.

In Europe, the North Sea contains western Europe's largest oil and natural gas reserves and is one of the world's key non-OPEC producing regions. Consequently, the region is a relatively high-cost producer. Five countries operate crude oil and natural gas production facilities in the North Sea: Denmark, Germany, the Netherlands, Norway, and the United Kingdom.

North America has enormous amounts of potential natural gas in underground reserves. Many sections of Canada have vast fields of natural

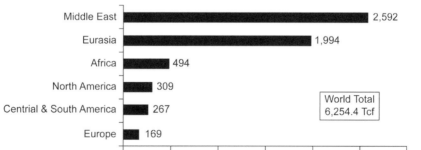

Much of the world's natural gas reserves are located in the Middle East in the countries of Qatar and Iran and in Eurasia, where Russia has the largest natural gas reserve. (*Source:* U.S. Department of Energy/Energy Information Administration)

Offshore drilling of gas and oil began in 1960 in Europe's North Sea. These oil and gas platforms are constructed on a foundation built on the bottom of the sea or are permanently anchored in place. (iStockphoto)

gas reserves and the country is a major gas producer in the western hemisphere. Most of the natural gas reserves in the United States are located around Texas and the Gulf of Mexico. The San Juan Basin is the second-largest deposit of natural gas in the United States. This reserve is found in the southwestern region of the United States, which includes New Mexico, Arizona, and parts of Colorado and Utah. The basin contains more than 15 percent of the nation's natural gas reserves.

Natural gas is produced in 32 U.S. states, but the top five states—Texas, Louisiana, Oklahoma, New Mexico, and Wyoming—produce 80 percent of the total. Altogether, the United States produces about 25 percent of the world's natural gas each year.

So, how much natural gas is available in the global reserves? Because natural gas is essentially irreplaceable, it is important to have an idea of how much natural gas is in the reserves. However, getting those numbers is not an easy task. In fact, no one really knows exactly how much natural

A worker checks pipes for the Portland Natural Gas Transmission System (PNGTS) in Wells, Maine, in 1998. PNGTS is a high-capacity, high-pressure, interstate natural gas pipeline that began serving New England's growing energy needs in 1999. (AP Photo/Robert F. Bukaty)

gas exists until it is extracted. Measuring natural gas in the ground is complicated, and it involves a great deal of inference and estimation. With new technologies, these estimates are becoming more and more reliable; however, they are still subject to revision.

WORLD GAS PRODUCTION COUNTRIES

Worldwide, natural gas is produced by two major coalitions of countries. The first is the Organization for Economic Cooperation and Development (OECD) and consists of three major groups: North America (United States, Canada, Mexico), much of Europe (Norway, Italy, France, and others), and Asia (Japan, South Korea, Australia/New Zealand). These member countries have pledged to work together to promote their economies, to extend aid to underdeveloped nations, and to contribute to the expansion of world trade.

The second group of major gas producers includes the non-OECD members. Some of the member countries include countries in the Middle East, Brazil, Africa, Russia, China, India, and parts of Europe. Let's look at the two major natural gas producers—Russia and Norway.

Russia

Russia holds the world's largest natural gas reserves, the second-largest coal reserves, and the eighth-largest oil reserves. Russia is also the world's largest exporter of natural gas, the second-largest oil exporter, and the third-largest energy consumer.

In 2007 Russia's real gross domestic product (GDP) grew by approximately 8.1 percent, surpassing average growth rates in all other countries and marking the country's seventh consecutive year of economic expansion. Russia's economic growth has been driven primarily by energy exports as a result of the increase in Russian oil production and the high world oil prices. Internally, in 2007 Russia received more than half of its domestic energy needs from natural gas, up from around 49 percent in 1992.

Norway

The vast majority of Norway's energy output is exported. In 2007 Norway was the world's third-largest net exporter of natural gas, behind Russia and Canada. This is largely attributable to the fact that almost all of Norway's electricity is generated by hydropower.

Norway's energy business sector is a major contributor to its economy. Exports of crude oil, natural gas, and refined petroleum products accounted for 68 percent of total exports, and this energy producing sector also contributes around one-third of country's budget revenue.

Norway's importance as a global energy supplier has grown steadily since production first began in the early 1970s. It is now the largest offshore oil producer in the world behind Saudi Arabia and Russia, and is the second-largest supplier of natural gas to continental Europe.

THE WORLD'S LARGEST CONSUMERS OF NATURAL GAS

The biggest consumers of natural gas in 2006 were the United States, Russia, Germany, and the United Kingdom. Additionally, the demand for natural gas in Spain grew by 92 percent from 2000, placing Spain in sixth place in Europe behind the United Kingdom, Germany, Italy, France, and the Netherlands.

United States

In 2005, according to the Energy Information Administration, International Energy Outlook 2009, natural gas accounted for 19 percent of net electricity generation in the United States. Coal-powered plants provided about 50 percent of the electricity demands for the country. In 2010 the natural gas share of electricity use in the United States was expected to reach about 21 percent. Natural gas supplies nearly one-fourth (23%) of all of the energy used in the United States. In 2004 the United States consumed 22.4 trillion cubic feet (Tcf) of natural gas. Because of its efficiency, cleanliness, and reliability, natural gas is growing increasingly popular. Consumption of natural gas will increase 20 percent by 2030, according to the U.S. Department of Energy (DOE). The top natural gas–consuming states in 2006 were Texas, California, Louisiana, New York, Illinois, and Florida.

Although in 2005, the United States imported more than 65 percent of the oil it used from abroad. However, in the same year, 97 percent of the natural gas used in the United States was produced in North America. The breakdown included 85 percent from the United States and 12 percent from Canada. Natural gas energy advocates report that every gallon equivalent of natural gas used in vehicles is one less gallon of petroleum that has to be imported.

Alaska North Slope

One of the United States' largest deposits of recoverable natural gas is located on Alaska's North Slope. Geologists estimate that this region contains an estimated 85.4 trillion cubic feet of undiscovered, technically recoverable gas from natural gas hydrates, according to a new assessment from the U.S. Geological Survey. According to current usage rates provided by the Energy Information Administration, this would be enough natural gas to heat more than 100 million average homes for 10 years. However, further research, including long-term production tests, still is needed to demonstrate gas hydrates as an economically producible resource.

Germany

Germany is the third-largest consumer of natural gas in the world, and Germany's natural gas reserves are the third-largest in the European

Union, after the Netherlands and the United Kingdom. Almost all of Germany's natural gas reserves and production occur in the northwestern areas of the country. Germany's sector of the North Sea also contains sizable natural gas reserves. However, environmental regulations have curtailed the complete exploration and development of the area. Despite the lack of domestic production, Germany is, as previously noted, the third-largest consumer of natural gas in the world, behind the United States and Russia.

United Kingdom

Britain's dependence on natural gas as a source of energy is growing, even as supplies from the North Sea are running out. With the depletion of gas from the United Kingdom continental shelf, Britain is becoming dependent on imports, either by pipelines from Norway or as liquefied natural gas from places farther away, such as Algeria and Qatar. By 2015 the United Kingdom is expected to import up to 80 percent of its gas supplies. Believe it or not, in 2004 the United Kingdom was a net exporter of natural gas.

France

France imports almost all of the natural gas it uses from Norway, Russia, the Netherlands, and Algeria. However, natural gas has a small share in France's energy portfolio. France's share of natural gas consumption in 2009 was estimated at approximately 10 percent.

Netherlands

The Netherlands depends on natural gas for about 60 percent of its domestic electricity. Coal is used for about 25 percent of its electricity needs. The Netherlands is also a major of exporter of natural gas.

Central and South America

Natural gas is the second-fastest growing energy source after nuclear power in Central and South America. For example, in Brazil, South America's largest economy, natural gas consumption in 2030 will increase by 30 percent from 2009.

Australia

Natural gas is the fastest-growing fuel in Australia and New Zealand and is expected to account for approximately 30 percent of the projected growth between 2005 and 2030. It also is expected be the fastest-growing fuel in the electric power sector. The country will be displacing more carbon-dioxide coal power plants with natural gas power generation.

LIQUEFIED PETROLEUM GAS—PROPANE

Another product that can be produced from natural gas is liquefied petroleum gas (LPG), commonly called propane. Propane is a gas that can be turned into a liquid and stored in pressure tanks. However, when propane is drawn from a tank, it changes to a gas. In many parts of the world, this gas is used as a fuel for cooking and heating or for motor vehicle fuel.

Liquefied petroleum gas is produced during natural gas processing and crude oil refining. In natural gas processing, LPG, butane, ethane, and pentane are removed prior to the natural gas entering the pipeline distribution system. About 55 percent of LPG processed in the United States is from natural gas. The other 45 percent comes from crude oil refining. Liquefied petroleum gas is the first product that results at the start of the crude oil refining process and is therefore always produced when crude oil is refined.

Liquefied petroleum gas has a long and varied history in transportation applications. It has been used in rural and farming settings since its inception as a motor vehicle fuel in 1912, and it is the third most commonly used fuel in the United States, behind gasoline and diesel. More than 350,000 light- to medium-duty vehicles running on LPG are used in the United States. Propane-fueled vehicles produce fewer emissions than do gasoline-fueled vehicles. As an example, propane-fueled vehicles produce 60 percent less carbon dioxide than gasoline-fueled vehicles.

NATURAL GAS VEHICLES

According to the Energy Information Agency (EIA), the greatest potential for large-scale substitution of natural gas for petroleum is in the transportation sector, especially in local fleet vehicles refueled at a central facility. Fleets generally operate a number of vehicles that are centrally maintained and fueled. They also travel more miles daily than the average personal use vehicle

Buses powered by compressed natural gas (CNG) are now common in the United States. These CNG buses belong to the fleet of the Los Angeles Metropolitan Transit Authority. (Publicimage/Dreamstime.com)

INTERVIEW

Green Advocate: Bob Walters, Technology Education Teacher, DeWitt Middle School, Ithaca, NY 14850
Go-Green Project: Designing Solar Sprint Model Cars

Describe your current position and your responsibilities in your particular program.

I am a technology education teacher and department head at the DeWitt Middle School in Ithaca, NY. I teach 8th grade and a computer class to 6th-grade students.

Where did you grow up, and what schools did you attend?

I grew up and attended the Massapequa High School, Massapequa, Long Island, New York.

What were some of your favorite activities and subjects when you attended high school?

I enjoyed the sciences. The typical science sequence was general science, biology, and earth science. Most people took earth science, and it had the reputation as the "science for dummies." Because I really enjoyed science, I took physics in 11th grade and chemistry in grade 12. Ironically, in the middle school in which I teach, many 8th-grade students accelerate to take 9th-grade earth science. Here it is an honors course.

In high school I wanted to take auto shop. That required engine mechanics first. In that class we learned the basics and rebuilt lawn mower engines. Then in auto shop we worked on teachers' cars. I was the "brake ace." I also rebuilt the carburetor on the family car and the transmission in my dad's van, the old parts of which have been a pencil holder on my desk for the last 33 years.

What colleges did you attend, and what was your major field?

I received a bachelor of science degree from SUNY Oswego, New York, in industrial arts and a master of arts degree from Indiana State University in School of Technology.

During this period, the industrial arts programs were starting to change, and the department was renamed technology education at the university; another grad student and I developed and taught a new course called communication technology. It was very popular.

What interested you in seeking a career in your discipline?

I just fell into this career. My high school guidance counselor told me that I was not "college material." That bothered me to the point that I brought my grades up.

Bob Walters and his students design and construct solar model cars to be used in the Junior Solar Sprint car competition. The students used a variety of materials to build their cars, including balsa wood, plastic soda bottles, and even a Pringles container. (Courtesy Bob Walters)

I went to a college that a friend was going to and found they had an industrial arts program. It looked fun, so I enrolled. The rest is history.

Describe your current model solar vehicle program.

For the most part, every student in New York State must take a school year's worth of technology education by the end of eighth grade. The course is hands-on, and students learn the design process, how to use tools and machines, and how to process various materials and information and are given design challenges. Much of the work is done in groups.

There are two complementary aspects of the Junior Solar Sprint (JSS) project. One is the JSS in-class activity, and the other is the multi-school competition.

For the Junior Solar Sprint class activity, students usually work in groups of three, to design and construct a vehicle. Each student is a lead engineer who designs one of the three subsystems, chassis, power train, or solar collection. The vehicle is built by the group. The design parameters are spelled out in the JSS rules available from the Northeast Sustainable Energy Association (NESEA): http://www.nesea.org/k-12/juniorsolarsprint/. Before they design or build, students take an open-book, online quiz to be sure they understand the design parameters.

The JSS multi-school competition has been held for 10 consecutive years. I added a design requirement to the rules. Student must submit sketches that show their ideation, part of the design process. Each student participant receives a commemorative t-shirt and has a chance to win trophies and medals.

The materials for constructing the vehicle largely come from the class supply budget. Other funds are needed for the t-shirts and awards. Some years we have sold pizza and water to raise money. However, the majority of money comes from grants. NESEA provides a grant. They are a pass-through agency. For the last several years, the money actually has come from the U.S. Army. Other money has also been secured through a variety of mini-grants. These have been available from various sources, including the professional association for technology teachers in New York State, local school district grants, the PTA, and the local power utility as well as the local universities.

How or why did you get interested in selecting this project?

I was always interested in the environment and saving energy. This seemed like a fun way to engage students in learning about green technology.

Explain the importance of the project as it relates to real-world issues.

Our country uses a disproportionate amount of energy and produces a disproportionate amount of greenhouse gases and pollution. To reverse this, change has to "start at home," beginning with new attitudes.

What materials and references (web sites, advisors, periodicals, etc.) did you use as resources for the project?

There are great resources available from the Northeast Sustainable Energy Association: http://www.nesea.org/k-12/. In addition to the JSS rules, there is a wealth of curriculum materials. There is a video that I show to students and have them take a quiz on, just to make sure they understand the challenge. The video is a bit dated but is available from the NESEA, and they even have the eight-minute introduction, which is all I show my students, online at http://www.nesea.org/k-12/juniorsolarsprint/modelsolarracecarteacherresources/.

There are many links at NYSERDA's site (http://www.getenergysmart.org/Default.aspx), including http://www.getenergysmart.org/EnergyEducation/Teachers/Curriculum.aspx.

Many of the materials for this project are available from Pitsco: http://www.pitsco.com/tabid/210/default.aspx?art=702. Some are available through Kelvin Electronics: http://www.kelvin.com/.

Although my own district is updating its web presence, and I have not been able to update my own site for some time, there are references that have been compiled there. These include samples of vehicles as well as subsystems of vehicles: http://www.icsd.k12.ny.us/dewitt/teched/jss.html.

What advice would you give other teachers who would like to do this project?

Although not required, if possible you can attend a JSS workshop. NESEA offers and coordinates these. Consider building some vehicles on your own and/or having a group of interested students build vehicles for fun or extra credit. Watch the video clip. The link is mentioned earlier in the interview.

Discuss some of the students' contributions to this project, and describe some of their reactions or comments.

Students come up with some very creative solutions to the problem. Most students really enjoy this project.

How long did it take to complete the project?

I spend about three to four weeks on this. Our classes meet each school day for 39 minutes.

Do you have any current plans to improve on or to extend this program?

This activity is also done in conjunction with other energy activities. Most recently, this has been a wind turbine activity: http://www.kidwind.org/. This was coordinated with the science teachers. They taught about societal impacts of wind energy, and the tech students designed and made wind turbine blades and conducted experiments of their own design. Other energy activities have involved fuel cell vehicles and energy conversion activities. This year I may start an energy club of some sort to provide students with opportunities to go further than they can in class.

and therefore can take better advantage of the lower price per gallon of natural gas. Among the fleets in which use of natural gas vehicles (NGVs) is already growing are taxi cabs, over-the-road trucks, street sweepers, transit buses, refuse haulers, school buses, delivery vehicles, airport shuttles, and forklifts.

As of 2010, there are more than 120,000 NGVs on U.S. roads and more than 8.7 million worldwide. Most NGVs are fueled at some 1,225 compressed natural gas stations throughout the United States, a number that has increased fourfold since 1991. Natural gas vehicles are the most commercially advanced of vehicles that are alternatively fueled (the others being those powered by methanol, ethanol, propane, and electricity).

The benefits of NGVs are most pronounced in congested urban areas that have air quality concerns, and fleet vehicles in those areas offer the most promise. Promotional federal initiatives include the Clean Air Act, the Clean Cities Program, the Congestion Mitigation Air Quality Program, the Energy Policy Act, and the Advanced Natural Gas Vehicle Program.

A number of factors make fleet vehicles—buses, taxis, and delivery vehicles—the prime target for natural gas. Because natural gas generally costs less than gasoline, these high-mileage vehicles can realize large savings in fuel costs. Also, fleet vehicles tend to be centrally located. Thus, fleets can locate near refueling stations, or they can install their own facility.

Natural gas vehicles offer tremendous benefits. Highway gasoline-powered vehicles account for roughly one-third of all carbon dioxide and nitrogen oxides emissions and half of all carbon monoxide emissions. Using natural gas rather than gasoline can produce major reductions in a number of vehicular emissions. In addition to being cleaner than conventional vehicles, NGVs reduce the nation's extreme dependence on imported oil, and the fuel cost is generally less than the cost of gasoline or diesel fuel.

But despite these benefits, and the fact that a survey of fleet operators by the Natural Gas Vehicle Coalition showed that NGVs are their favorite type of alternatively fueled vehicle, NGVs face serious hurdles. Market growth for these vehicles has not been dramatic. The primary obstacle is that vehicle production levels are limited, making the purchase price of an NGV higher than that of a comparable conventionally fueled vehicle. Even for high-mileage vehicles, it is difficult to offset the extra thousands of dollars on an NGV's price tag with fuel-cost savings of 10–20 cents per gallon. Once the demand for NGVs reaches a level that can sustain full production, prices will fall. Currently, the natural gas fueling infrastructure also is limited, but the fueling infrastructure will expand as NGVs gain in popularity.

Natural gas costs, on average, one-third less than conventional gasoline at the pump. More than 50 different manufacturers produce 150 models of light-, medium-, and heavy-duty vehicles and engines. Roughly 22 percent of all new transit bus orders are for natural gas. Natural gas is sold in gasoline gallon equivalents, or GGEs. A GGE has the same energy content (124,800 Btus) as a gallon of gasoline.

Drawbacks of Natural Gas for Vehicles

The United States would need a lot more natural gas stations to power a third of its vehicles. Natural gas is still a fossil fuel: it might be cleaner-burning than oil, but it is still a hydrocarbon that has to be taken out of wells and is in limited supply. Natural gas vehicles have a shorter driving range than regular gas-powered vehicles because natural gas has a lower energy content compared to gas.

Benefits of Natural Gas Vehicles

Exhaust emissions from a typical NGV are much lower than those from gasoline-powered vehicles. In addition, dedicated NGVs produce little or no evaporative emissions during fueling and use. In gasoline vehicles, evaporative and fueling emissions account for at least 50 percent of a vehicle's total hydrocarbon emissions.

Typical dedicated NGVs can reduce the following exhaust emissions:

FEATURE

Using Natural Gas to Power Motor Vehicles

Many taxi and bus drivers in Cairo, Egypt, have converted their gasoline-powered engines to run on natural gas. Cairo is the world leader in the number of privately owned natural gas–powered motor vehicles, and now the country's bus and taxi companies are coming aboard. Egypt has abundant natural gas reserves and can offer car owners a fuel that is less expensive than gasoline and a cleaner-burning fuel than gasoline. Natural gas vehicles produce about 80 percent less carbon monoxide and fewer hydrocarbons than gasoline-powered vehicles. And natural gas costs less than gasoline in Egypt. A cubic meter of natural gas is 50 percent less expensive than the equivalent amount of gasoline. By 2010 Egypt is expected to have more than 25 stations to service natural vehicles.

- Carbon monoxide by 70 percent
- Non-methane organic gas by 87 percent
- Nitrogen oxides by 87 percent
- Carbon dioxide (CO_2) by almost 20 percent below those of gasoline vehicles

Natural gas vehicles also produce far less urban emissions than diesel vehicles. The NGVs produce less amounts of nitrogen oxides than comparable diesel engines.

Natural gas contains less carbon than any other fossil fuel and thus produces lower carbon dioxide emissions per vehicle mile traveled. Although NGVs do emit methane, another principle greenhouse gas, any increase in methane emissions is more than offset by a substantial reduction in CO_2 emissions compared to other fuels. Tests have shown that NGVs produce up to 20 percent less greenhouse gas emissions than comparable gasoline vehicles and up to 15 percent less than comparable diesel vehicles.

Are Natural Gas Vehicles Safe?

The fuel in NGVs, unlike gasoline, dissipates into the atmosphere in the event of an accident. On the other hand, gasoline pools on the ground create a fire hazard. The fuel storage cylinders used in NGVs are much stronger than gasoline fuel tanks. Natural gas vehicle cylinder designs are subjected to a number of federally required "severe abuse" tests, such as heat and pressure extremes, gunfire, collisions, and fires.

Natural gas vehicle fuel systems are sealed, which prevents any spills or evaporative losses. Even if a leak were to occur in an NGV fuel system, however, the natural gas would dissipate into the air because it is lighter than air.

Natural gas has a high ignition temperature, about 1,200 degrees Fahrenheit, compared with about 600 degrees Fahrenheit for gasoline. It also has a narrow range of flammability; that is, in concentrations in air below about 5 percent and above about 15 percent, natural gas will not burn. The high ignition temperature and limited flammability range make accidental ignition or combustion of natural gas unlikely. Natural gas is not toxic or corrosive and will not contaminate ground water.

A gas station attendant refuels a vehicle with compressed natural gas at one of Cairo, Egypt's several fueling stations. As of January 1996, almost all taxi drivers in Cairo, under a government project, have converted their engines to work with natural gas. Natural gas is half the price of gasoline and offers a means of fighting air pollution cheaply and safely. (AP Photo/Leila Gorchev)

NATURAL GAS BENEFITS

Natural gas (largely methane) burns cleaner than the other fossil fuels (45% less carbon dioxide emitted than coal and 30% less than oil). It is easily transported via pipelines and fairly easily using tankers (land and sea). It can be piped into homes to provide heating and cooking and to run a variety of appliances. Where homes are not piped, it can be supplied in small tanks. It can be used as a fuel for vehicles (cars, trucks, and jet engines), where it is cleaner than gasoline or diesel.

NATURAL GAS EMISSIONS

Pollutants emitted in the United States, particularly from the combustion of fossil fuels, have led to the development of many pressing environmental problems. Natural gas, emitting fewer harmful chemicals into the atmosphere than other fossil fuels, can help to mitigate some of these environmental issues. Natural gas is an extremely important source of energy for reducing pollution and maintaining a clean and healthy environment. In addition to being a domestically abundant and secure source of energy, the use of natural gas also offers a number of environmental benefits over other sources of energy, particularly other fossil fuels.

Natural gas is the cleanest of all the fossil fuels. Because carbon dioxide makes up such a high proportion of U.S. greenhouse gas emissions, reducing carbon dioxide emissions can play a huge role in combating the greenhouse effect and global warming.

As noted previously, the combustion of natural gas emits almost 30 percent less carbon dioxide than oil and just under 45 percent less carbon dioxide than coal. Composed primarily of methane, the main products of the combustion of natural gas are carbon dioxide and water vapor, the same compounds we exhale when we breathe.

Coal and oil are composed of much more complex molecules, with a higher carbon ratio and higher nitrogen and sulfur contents. This means that when combusted, coal and oil release higher levels of harmful emissions, including a higher ratio of carbon emissions, nitrogen oxides, and sulfur dioxide. Coal and fuel oil also release ash particles into the environment, substances that do not burn but instead are carried into the atmosphere and contribute to pollution. The combustion of natural gas, on the other hand, releases very small amounts of sulfur dioxide and nitrogen oxides, virtually no ash or particulate matter, and lower levels of carbon dioxide, carbon monoxide, and other harmful emissions.

FUTURE OF NATURAL GAS

According to government studies, worldwide natural gas consumption will increase from about 100 trillion cubic feet in 2005 to 158 trillion cubic feet in 2030. Natural gas will probably replace petroleum and coal wherever

 **VIDEO**

U.S. The Natural Gas Star Program: Partnered with the Environmental Protection Agency, this video claims to be a blueprint for resource management. For more, go to http://www.epa.gov/gasstar/documents/videos/processing.html (06:32 minutes).

Carbon Dioxide Emissions per Vehicle Mile Traveled in Vehicle and Non-Vehicle Stages of the Fuel Cycle for Various Fuels

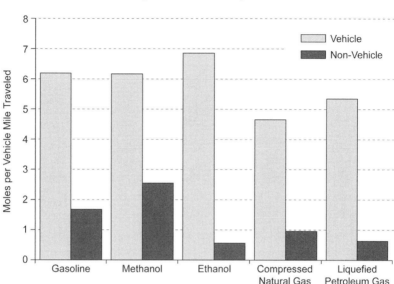

(*Source:* U.S. Department of Energy/Energy Information Administration)

possible. The reason is that natural gas combustion produces less carbon dioxide than coal or petroleum production and products. Therefore, natural gas is expected to remain a key energy source for the industrial sector. The industrial sector, made up of such major manufacturing groups as plastics, chemical, and steel making, will be large consumers of natural gas, accounting for 43 percent of the projected natural gas use in 2030. Natural gas will also be a major supplier of electricity. Electricity generation will account for 35 percent of the world's total natural gas consumption in 2030.

The Gas Resource of the Future—Methane Hydrate?

Historically, the United States has produced much of the natural gas it has consumed, with the balance imported from Canada through pipelines. According to Energy Information Administration, total U.S. natural gas consumption is expected to increase from about 23 trillion cubic feet in 2008 to 26 trillion cubic feet in 2030—a projected jump of more than 18 percent.

However, production of domestic conventional and unconventional natural gas cannot keep pace with demand growth. So the development of new, cost-effective resources such as methane hydrate, huge amounts of which underlie the Arctic polar regions and ocean sediments, can play a major role in moderating price increases and ensuring adequate future supplies of natural gas for American consumers.

Methane hydrate is a crystalline combination of natural gas molecules and water molecules. The caged-like molecules look like ice but burn if they are lighted with a match. Inside the "ice" are trapped molecules of methane, the chief product of natural gas. If methane hydrate is either warmed or depressurized, it will revert back to water and natural gas. When brought to Earth's surface, one cubic yard of gas hydrate releases approximately 450 cubic yards of natural gas.

Hydrate deposits may be several hundred meters thick and generally occur in two types of settings: under Arctic permafrost and beneath the ocean floor. Methane hydrate is stable in ocean floor sediments at water depths greater than 930 feet. Estimates on how much energy is stored in methane hydrates range from 350 years' worth to 3,500 years' worth.

More Work Ahead and Concerns about Methane Hydrates

Although methane hydrates have the potential to offer a clean source of energy, more research needs to be conducted. For example, because methane

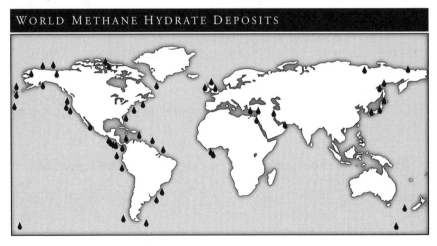

Discoveries of methane hydrate deposits, first in polar regions and then throughout the deep-water shelves of every continent, reveal that natural methane hydrate occurs on a truly staggering scale. (*Source:* National Oceanic and Atmospheric Administration)

is also a greenhouse gas, release of even a small percentage of total deposits could have a serious effect on Earth's atmosphere. Despite its short atmospheric half-life of seven years, methane has a global warming potential. Another concern is that unstable hydrate layers could give way beneath oil platforms, which could cause the freed gas to explode.

Landfill Gas from Biomass

Scientists are also researching ways to produce natural gas (methane) from biomass. Biomass is a term used to describe the total amount of living matter in a particular area at any given time. The energy from biomass is the oldest fuel used by humans.

Drilling in Landfills to Recover Methane

In 2009 more than four billion cubic feet of landfill methane gas was used for heating and electricity production. Landfill gas is created when microorganisms cause organic waste, such as food wastes and paper, to decompose in landfills. Landfill gas is composed of about 50 percent methane. Carbon dioxide and volatile organic compounds (VOCs) make up the remainder.

Landfill gas escapes into the air unless it is collected and burned. In landfill gas energy projects, landfill gas is burned in boilers, special engines, and combustion turbines to produce electricity. The landfill's size and age, the quantity of organic waste, and the local climate help determine how much gas a landfill can produce. The Environmental Protection Agency (EPA) requires large landfills to collect and burn landfill gas with flares to destroy the VOCs.

Although some landfills simply burn landfill gas with a flare, as of 2010 more than 380 projects at 365 U.S. landfills are collecting and using landfill gas to produce energy. The EPA estimates that more than 600 additional landfills could support landfill gas energy projects cost-effectively.

Landfill gas continues to be produced for 20 years or more after a landfill is closed. Therefore, as long as landfills continue to be built, landfill gas will continue to be a resource for producing electricity.

Some Benefits

Burning landfill gas to produce electricity has little impact on land resources. Although the equipment used to burn the landfill gas and generate electricity does require space, it can be located on land already occupied by the existing landfill, thus avoiding any additional use of land.

Air Emissions

Burning landfill gas produces nitrogen oxide emissions as well as trace amounts of toxic materials. The amount of these emissions can vary widely, depending on the waste from which the landfill gas was created. The carbon dioxide released from burning landfill gas is considered to be a part of the natural carbon cycle of the earth. Producing electricity from landfill gas avoids the need to use nonrenewable resources to produce the same amount of electricity. In addition, burning landfill gas prevents the release of methane, a greenhouse gas, into the atmosphere.

In Europe, processing plants use up to 50 percent of municipal trash for energy production. Energy trash processing plants are also located in several American cities in Maryland, California, Illinois, Ohio, Wisconsin, and Washington.

Using Marine Plants to Produce Methane

Another source of methane may be marine plants, according to a recent British report. Algae, as well as other marine biomass such as kelp, could have an important role in the future of energy production. The methane from this biomass could be used to generate electricity and heat or used as compressed natural gas for transportation fuel.

In 2009 a team of marine scientists from the United Kingdom and Ireland received funding to determine the possibility of producing renewable fuel using seaweed, through a study of the brown seaweed *Laminaria hyperborean* more commonly referred to as kelp. Their goal is to determine whether large areas of kelp grown through ocean farming could be used as biomass for the production of methane and ethanol. There are approximately 386 square miles of kelp forests in the waters around the United Kingdom that are dense enough to be harvested.

More research is necessary to find better methods for the conversion step, biomass to methane, on a large scale, but the work already done shows that methane energy can result from algae biomass.

DID YOU KNOW?

Kelp grows one and a half feet per day.

BOOKS AND OTHER READING MATERIAL

Burleson, Clyde W. *Deep Challenge!: The True Epic Story of Our Quest for Energy Beneath the Sea.* Burlington, MA: Gulf Professional Publishing, 1998.

Natural Gas Supply Association. *Natural Gas and the Environment.* www.naturalgas.org/environment/naturalgas.asp

Sietz, John L. *Global Issues: An Introduction.* Malden, MA: Blackwell, 2002.

Snedden, Robert. *Energy from Fossil Fuels.* Boston: Heinemann Library, 2001.

U.S. Department of Energy, Energy Information Administration. http://www.eia.doe.gov.

U.S. Department of Energy, Office of Fossil Energy. http://www.fe.doe.gov.

U.S. Geological Survey Energy Resources Program. http://energy.usgs.gov/index.html.

SOMETHING TO DO

1. On a world map locate the 10 largest producers of natural gas, explain the infrastructure needed to transport and distribute the gas over long distances, and identify and analyze the political "chokepoints" that could disrupt the distribution of the gas.

2. Marcellus shale covers most of New York State. Located deep within the Marcellus shale formations are large deposits of natural gas. Extracting the gas is a delicate process. Research how it might be accomplished and the dangers to human and animal inhabitants.

3. Conduct research to discover how we get natural gas, how it is stored and delivered, and how it impacts the environment. For references and additional research, visit the following web sites: www.eia.doe.gov; www.loe.org; www.citizenscampaign.org.

4. Go to the Oil and Gas Journal Online Web site, read a selection of the articles, and summarize what you determine to be the journal's purpose and audience.

WEB SITES

The following Web sites, although not inclusive, include government and nongovernmental organizations.

www.aga.org
> The American Gas Association, founded in 1918, represents 195 local energy companies that deliver clean natural gas throughout the United States.

www.ngsa.org
> The National Gas Supply Association (NGSA) represents suppliers that produce and market natural gas. Established in 1965, NGSA encourages the use of natural gas within a balanced national energy policy.

http://www.iangv.org
> The International Association for Natural Gas Vehicles was established in 1986 to provide the NGV industry with an international forum and to foster growth, safety, product development, and policy formation. In June 2010 the Association was renamed to NGV Global.

http://www.energy.gov/energysources/naturalgas.htm
> The Department of Energy Office of Fossil Energy invests in research and development of technologies in the areas of natural gas supply, delivery reliability, and utilization. Through the Strategic Center for Natural Gas, DOE works with industry to develop technologies to support this fuel.

VIDEOS

The following video and audio selections are suggested to enhance your understanding of energy topics and issues. The author has made a consistent effort to include up-to-date Web sites. However, over time, some Web sites may move or no longer be available.

Viewing some of these videos may require special software called plug-ins. Therefore, you may need to download that software to view the videos. You also may need to upgrade your player to the most current version.

> **U.S. Texas (Issues):** This infrared video reveals that "clean" natural gas is not always what it seems. Follow this helicopter video of fugitive emissions: http://txsharon.blogspot.com/2009/08/clean-burning-natural-gas-has-dirty.html (03:15 minutes).

U.S. Natural Gas Star Program: Created in partnership with the Environmental Protection Agency, this video claims to be a blueprint for resource management: http://www.epa.gov/gasstar/documents/videos/processing.html (06:32 minutes).

South America—Natural Gas in Columbia: This short film documents a project to connect poor homes with natural gas, to improve the lives of those in this impoverished section of Columbia: http://www.youtube.com/watch?v=watcK8hi5RA (13:28 minutes).

General (Global Uses): Do you know the fundamental difference between propane and butane? To discover how one carbon atom makes the difference, go to http://www.ehow.com/video_4756915_what-difference-between-propane-butane.html (01:29 minutes).

Chapter 4

Coal

The world's largest producers and consumers of coal are China, Poland, Russia, India, and the United States. In the United States, coal accounts for approximately 49 percent of electricity output. Wyoming is the largest coal-producing state and Texas is the largest coal-consuming state and it is also the largest consumer of electricity.

Coal has been used for thousands of years. There is archeological evidence that China was burning coal in 1100 B.C. But not until brick chimneys became popular did people burn coal indoors. During the Industrial Revolution in England, the common use of steam engines led to a surge in the demand for coal. As an example, approximately 100,000 coal-fed steam engines were used to power machinery, trains, and steamboats and for pumping water out of coal mines.

COAL: A MAJOR SOURCE OF THE WORLD'S ENERGY

According to a study by International Energy Outlook, coal's share of world energy consumption in 2006 is projected to increase by 48 percent in 2030. Coal's share of the electric power sector will reach 46 percent by 2030.

HISTORY OF COAL

During the Industrial Revolution in the 18th and 19th centuries, the demand for coal surged. The major reason for the growth in the use of coal was improvement of the steam engine. The steam engine was designed by James Watt and patented in 1769, and Watt used coal to make the steam to run his engine. Steamships and steam-powered railroads were becoming the chief forms of transportation. All of these vehicles and boats used coal to fuel their boilers for mechanical power.

With the development of electric power in the 19th century, coal's future became closely tied to electricity generation. Thomas Edison developed the first practical coal-fired electricity-generating power plant. The coal-fired power plant went into operation in New York City in 1882, supplying electricity for household lights.

In the second half of the 1800s, more uses for coal were found. By 1875, coke (which is made from coal) had replaced charcoal as the primary fuel for iron blast furnaces to make steel.

MAJOR USES OF COAL

Today, coal's primary use is for the generation of electricity. In the United States, coal generates approximately 50 percent of the electricity consumed in America each day, far more than any other energy source. Worldwide, coal generates 40 percent of total electricity. Coal is also used in retail businesses and the industrial sector as a fuel for heating and for powering steel-making plants, cement plants, and other industrial and manufacturing facilities.

Other important users of coal include alumina refineries, paper manufacturers, and the chemical and pharmaceutical industries. Several chemical products can be produced from the by-products of coal. For example, refined coal tar is used in the manufacture of chemicals such as creosote oil,

DID YOU KNOW?

In North America, during the 1300s in what is now the U.S. southwest, the Hopi Indians used coal for cooking, for heating, and to bake the pottery they made from clay.

> **Products Made From Coal**
> - insecticides
> - paint thinners
> - varnishes
> - insulation
> - fuel
> - perfumes
> - medicines
> - fertilizers
> - linoleum
> - paint pigments
> - food preservatives
> - batteries

naphthalene, phenol, and benzene, and ammonia gas recovered from coke ovens is used to manufacture ammonia salts, nitric acid, and agricultural fertilizers.

Thousands of different products have coal or coal by-products as components: soap, aspirins, solvents, dyes, plastics, and fibers such as rayon and nylon. Coal is also an essential ingredient in the production of specialty products, of which the following are examples:

- activated carbon—used in filters for water and air purification and in kidney dialysis machines
- carbon fiber—an extremely strong but lightweight reinforcement material used in construction, mountain bikes, and tennis rackets
- silicon metal—used to produce silicones, which are in turn used to make lubricants, water repellents, resins, cosmetics, hair shampoos, and toothpastes

WHAT IS COAL AND HOW IS IT FORMED?

Coal is a blackish organic substance and the most abundant fossil fuel; it is used primarily to produce electricity and to a lesser degree to heat buildings. Environmental issues associated with coal include air pollution from coal-fired power plants and the impact of coal mining on natural resources.

Coal is a mixture of carbon and various other materials formed from the accumulation of partially decayed plants in large, shallow swamps, lakes, and marshes millions of years ago. It is found in beds and seams both near the surface and underground. The transformation of organic deposits into coal involved compaction and compression by burial under hundreds and hundreds of feet of sediments. The formation of peat was the first step in the coal-making process. Over time, the peat was compacted beneath other deposits. As a result, water was squeezed out of the peat, and gases such as methane were expelled into the atmosphere. Over thousands of years, the continued burial and compression caused the peat to alter into different grades of coal: lignite, bituminous, and anthracite. Lignite is the lowest grade, with the highest percentage of volatile matter. Bituminous, the next grade from lignite, is the most abundant of the three types of coal. Anthracite, or hard coal, is the highest-grade coal, with high carbon content and a low percentage of volatile matter.

WHERE IS COAL FOUND?

Coal deposits are found all over the world. Even Antarctica has coal deposits. However, most of the coal reserves are found in large deposits in the midlatitudes of the Northern Hemisphere. There are fewer coal deposits in the Southern Hemisphere. In all, about 100 countries have coal reserves.

Most of the world's largest deposits are in North America, eastern Europe, Russia, China, India, and Africa. In the United States, coal is found in 38 states, and nearly one-eighth of the country lies over coal beds. Some of the top coal-mining states include: Montana, Illinois, Wyoming, West Virginia, Kentucky, Pennsylvania, Ohio, Colorado, Texas, and Indiana.

HOW IS COAL MINED?

There are two basic ways to mine coal. Surface mining is used when coal is found close to the surface or on hillsides. Underground mining, or subsurface mining, is used to extract coal deep beneath Earth's surface or in coal seams on hillsides.

How Coal is Formed

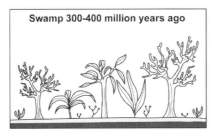

Before the dinosaurs, many giant plants died in swamps.

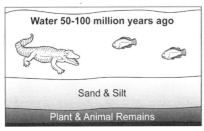

Over millions of years, the plants were buried under water and dirt.

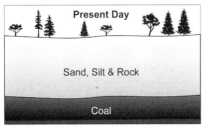

Heat and pressure turned the dead plants into coal.

(*Source:* U.S. Department of Energy/U.S. Energy Information Administration/ Illustrator: Jeff Dixon)

Surface Mining and Reclamation

Surface mining is essentially a process that involves removing the overburden (earth and rock covering the coal) with heavy earth-moving equipment, scooping out the coal, replacing the excavated soil, and reestablishing vegetation and plant life, a process known as reclamation.

Much of the reclamation activity in the United States was the result of legislation passed in 1977 called the Surface Mining Control and Reclamation Act. The act requires companies involved in surface mining operations to restore mined lands back to their natural conditions after the mining operations cease. It also prohibits surface mining on certain lands, such as national forests. Reclaimed land has been successfully used for wildlife preserves, golf courses, recreational parks, pasture land, native habitats, productive farmland, and for commercial development.

The advantages of surface mining are many. Where it can be used, it permits recovery of 90 percent or more of the coal to be mined.

To remove coal from regions located at or near Earth's surface, large equipment such as a bulldozer is generally used. Bulldozers remove soil and rock to expose coal, which is located near the surface. Surface mining may involve digging up approximately between 30 and 80 feet of topsoil and rock to reach the underground coal layers. Compared to underground mining, surface mining generally costs less, is safer for miners, and usually results in the removal of a greater percentage of the coal underground. However, it also results in extensive disruption of the land.

Surface mining can cause environmental problems. Common problems include the destruction of ecosystems and habitats. In addition, the removal of vegetation involved in surface mining makes an area more prone to soil erosion and landslides. Water pollution can also occur in nearby streams. As water leaches through tailings leftover from mining operations, it picks up minerals and carries them into groundwater reserves or to lakes and streams. Acid mine drainage is also a pollution problem that results from many types of surface mining. In many countries, including the United States, government agencies enforce reclamation operations that restore the land to its pre-environmental conditions once the mining operations cease.

Underground Mining

Underground mining is used to extract coal lying deep beneath the Earth's surface or in seams exposed on hillsides. The coal is reached through the drilling of two openings into the coal bed to transport workers and equipment and to send coal to the surface. Both openings serve to circulate air in the mine. Coal is then broken up and mined by one of several methods, including the following:

Conventional Mining: Conventional mining is an older practice of using explosives to break up coal seams.

Continuous Mining: In continuous mining, a huge machine with a large rotating steel drum equipped with tungsten carbide teeth scrapes coal from a seam at high speeds.

Longwall Mining: In longwall mining, a cutting machine with a large rotating steel drum is dragged back and forth across a "long wall" or seam of coal. The loosened coal falls onto a conveyer belt for removal

There are several forms of surface mining; one is the removal of shallow coal over a broad area where the land is mostly flat. (Airphoto/Dreamstime.com)

from the work area. After coal is extracted, it is removed by automatic extraction systems that cut the coal, which is loaded onto shuttle cars in a central loading area in the mine or placed on belt conveyors, which remove coal to the surface.

Environmental Issues in the Mining of Coal

Acid Mine Drainage

Mining coal can result in environmental problems, including the destruction of ecosystems and habitats. In addition, the removal of vegetation involved in surface mining makes an area more prone to soil erosion and

Coal is the official state rock of Utah.

landslides. Water pollution can also occur when runoff of wastes enters nearby streams. Water leaching through mine tailings leftover from mining operations picks up minerals and carries them into groundwater reserves or into lakes and streams.

Underground mines and open-pit mines can be a serious environmental problem if left abandoned. Acid mine drainage (AMD) is a water pollution problem resulting from the discharge into streams or rivers of acidic water from coal or other mines containing iron, copper, lead, or zinc mineral ores. Acid mine drainage also results when rainwater leaches through overburden or tailings—the waste materials produced by mining operations. Such water leaching through mine shafts and tailings causes chemical reactions to occur. The combination of air, dissolved oxygen in the water, and the activities of organisms that synthesize nutrients from inorganic chemicals causes iron-sulfide compounds in ores and waste rock to oxidize, producing a high concentration of sulfuric acid (H_2SO_4). When released into streams, the acidic solution is toxic to aquatic life. The acid can also leach into and pollute groundwater.

Acid mine drainage is a potential problem in any area with abandoned coal or metal mines or where deposits of mine tailings are present. Some states with AMD problems are Pennsylvania, West Virginia, Colorado,

J. Scott Horrell, the environmental program manager with the Department of Environmental Protection's (DEP) Bureau of Abandoned Mine Reclamation, shows acid mine drainage flowing from an entrance to an old mine in Fallston, Pennsylvania, in May 2005. DEP officials planned to seal several entries to the mine, which was abandoned in the 1960s. (AP Photo/Keith Srakocic)

Ohio, Wyoming, and Oklahoma. Outside the United States, AMD has been reported in Indonesia and South Africa. Although not yet reported elsewhere, AMD is likely a problem in other countries that now have, or once had, heavy coal, zinc, iron, copper, or lead mining industries.

Reducing Acid Mine Drainage Problems

Acid mine drainage pollution can be reduced. One method involves sealing abandoned mines to prevent water from flowing in or out, thus eliminating the discharge of acidic water into streams. Chemical treatment, in which limestone or lime is used to neutralize acids that form in mines, is most often used to eliminate AMD. Another successful method for reducing AMD involves using natural and human-made bog-type wetlands to filter sulfuric acid from mine wastewater before it enters streams and rivers. Organic matter, bacteria, and algae all work together to filter, absorb, and precipitate out the heavy metal ions and raise the pH level. More than 300 wetland water treatment systems have been built in the United States, many in coal-mining regions.

The U.S. Environmental Protection Agency (EPA) has established regulations to limit acid levels of mine drainage to no net acidity. The regulations require the pH of discharge to be between 6.0 and 9.0. The average total iron content of the discharge must be less than 3 mg/L. According to the EPA standards, new mines must be designed and operated to meet the standard of zero discharge.

Cleanup Techniques for Acid Mine Drainage

Ohio University's Mary Stoertz, a professor of geology, and others have developed a unique way to clean up AMD resulting from tailings surrounding abandoned coal mines located in southeastern Ohio. A major stream in the area of the study was highly acidic because of rainwater that leached sulfur from coal mine tailings. The cleanup organization used flue gas desulphurization (FGD), an alkaline substance, to neutralize the acid in the stream. Flue gas desulphurization is a waste byproduct formed when calcium carbonate ($CaCO_3$) is used by power plants to scrub sulfur dioxide (SO_2) from smokestack gases. A local power plant donated the FGD to the university research team. There are also other experimental trials in Pennsylvania using fly ash mixed with a solidifier to fill abandoned coal mines.

Coal Ash Slurry Spills

Coal power plants burn a lot of coal per year. Besides producing electricity, these power plants also produce millions of tons of coal ash waste. The coal ash wastes are stored in special lagoons or containment ponds in which a mixture of coal ash and water produces coal ash slurry, which resembles a thick mud or sludge. There are hundreds of these lagoons or ponds containing ash waste slurry throughout some of the leading coal-mining regions in the United States.

At times, the toxic material in the slurry has leached out of some of these containment ponds, causing health problems for humans and wildlife. The toxic material can penetrate into groundwater reservoirs and local streams as well. A major concern is that some of the dams containing the slurry could break apart, causing millions of liters of slurry to spill out into the environment, killing wildlife and damaging homes and businesses. Now the U.S. government is planning policies to regulate and govern coal ash waste disposal sites in those coal regions that have coal ash slurry containment ponds or lagoons.

HOW IS COAL TRANSPORTED?

Once the mining of coal is completed, the coal is ready to be shipped. Shipping coal to various places in the country and overseas is a major business. Most coal sent from mining operations to various geographical areas travels either by rail or by barges. Trucks and covered conveyor systems are used to move coal over shorter distances. There is even a coal slurry pipeline (which mixes coal with water and sends it through a metal tube to its destination) connecting a mine in Arizona with a power plant in Nevada that handles several million tons annually. Lake carriers and ocean vessels move huge quantities of coal shipments across the Great Lakes and to countries overseas. Much of this coal is used to produce electricity.

DID YOU KNOW?

In 2008, in Kingston, Tennessee, approximately one billion gallons of coal ash slurry and sludge spilled out of a containment pond and covered more than 400 acres of land and contaminated local water resources.

A Tennessee Valley Authority towboat heads off after pushing a bargeload of coal into place to be unloaded at the Cumberland City Fossil plant in Cumberland City, Tennessee. The plant uses several thousand tons of coal each day. (AP Photo/The Leaf-Chronicle, Greg Williamson)

HOW DOES A COAL-FIRED POWER STATION PRODUCE ELECTRICITY?

Fossil fuel power plants using coal or natural gas convert the energy stored in fossil fuels such as coal, oil, or natural gas successively into thermal energy, mechanical energy, and finally electric energy. The first step toward using coal as an energy source is to pulverize the coal into fine powder for proper combustion. The coal is then fed into a boiler to convert water into steam. The steam is then used to drive steam turbines connected to a generator shaft, which generates electricity. Then the electricity is distributed across a wide geographic area through transmission lines.

 DID YOU KNOW?

The U.S. coal industry currently employs 80,000 people, down from 700,000 in the 1920s, when production was half of what it is today.

Peabody Energy: What Is Coal? To learn more about Earth's coal, go to http://www.schooltube.com/video/39882/Peabody-Energy-What-Is-Coal (02:33 minutes).

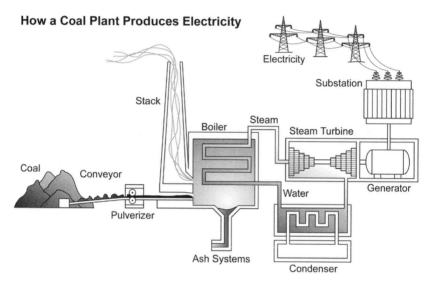

Almost 50 percent of all electricity produced in the United States is generated by coal-fired plants. (Illustrator: Jeff Dixon)

MAJOR COAL-PRODUCING STATES

The United States is the world's second-largest coal producer after China, and its coal production is spread throughout the country.

In the United States, coal adds $81 billion a year to the economy and more than 1 million jobs for Americans. As stated earlier, in the United States, coal is the major power behind our electricity needs.

More coal is produced in the state of Wyoming than in any other state in the United States. One of the reasons is the low sulfur content of the coal found in the Powder River Basin. Every time you turn on a light or turn on your computer, it is possible that you are using coal, maybe even Powder River coal. The Powder River Basin in Wyoming accounts for 35 percent of all the nation's coal production and produces low-ash, low-sulfur coal suitable for use since the Clean Air Act of 1990. West Virginia is responsible for about 14 percent of all coal production in the United States, followed by Kentucky, where about 10 percent of U.S. coal production takes place.

DID YOU KNOW?

Coal mining in Pennsylvania fueled the Industrial Revolution in the United States in the mid-1700s.

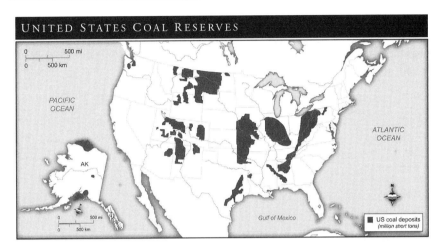

Wyoming is the largest producer of coal in the United States and has the largest number of surface coal mining operations. Some of the country's largest underground mining operations are in West Virginia, Pennsylvania, Ohio, and Colorado. (*Source:* U.S. Department of Energy/Energy Information Administration, Coal Reserves 2007)

MAJOR GLOBAL COAL-PRODUCING COUNTRIES

Coal is the world's most abundant and widely distributed fossil fuel. According to the International Energy Agency, by 2030 coal-based power generation will provide about one-third of all global electricity generation.

Major hard-coal producers include China, the United States, India, Australia, South Africa, Russia, Indonesia, Poland, Ukraine, and Kazakhstan. Although coal deposits are widely dispersed, more than 59 percent of the world's recoverable reserves are located in five countries—Australia, China, India, the United States, and Canada.

Germany

Germany ranks seventh worldwide in coal production and ranks first in Europe and fourth worldwide in coal consumption. If the last coal-mining pits are closed by 2018 as planned, it will mark an end to a long chapter in

German coal history. However, the country might have to import coal for its power plants even though its mines will be closed.

India

India has a long history of commercial coal mining starting from 1774. India relies mostly on coal to meet the nation's energy demands, which are growing along with its economy. Given India's commitment to reduce its coal-fired emissions by 25 percent by 2030, the government has set up plans and programs to search for alternative fuels.

China

According to the U.S. Energy Information Administration (EIA), China is the largest producer and consumer of coal in the world, and many of China's large coal reserves have yet to be developed. As of 2006, coal accounts for about 70 percent of China's total energy consumption.

China is building coal plants that first gasify the coal before it burns. This process allows coal plants to burn coal between 30 percent and 44 percent more efficiently. However, coal will remain a critical part of China's energy mix for decades to come, but growth will slow and then peak at perhaps 3.4 billion tons per year by 2020, according to China's Energy Research Institute.

Poland

Poland is one of the leading coal producers in the world with an annual coal production of more than 160 million metric tons. Poland has three major Carboniferous coal basins, namely the Lower Silesian, Upper Silesian, and Lublin basins.

EXPORTERS OF COAL

Australia

According to the Australian Coal Industry, Australia is the world's biggest coal exporter, and black coal is Australia's largest export, worth more than $A50 billion in 2008–2009. With extensive coal reserves concentrated along the nation's eastern seaboard in New South Wales and Queensland, Australia has more than 76 billion tons of identified black coal reserves, which will last for more than 200 years at current rates of production.

Top Coal Exporters (2007)

	Total of which	Steam	Coking
Australia	244Mt	112Mt	132Mt
Indonesia	202Mt	171Mt	31Mt
Russia	100Mt	85Mt	15Mt
Colombia	67Mt	67Mt	-
South Africa	67Mt	66Mt	1Mt
PR China	54Mt	51Mt	3Mt
USA	53Mt	24Mt	29Mt

(*Source:* U.S. Department of Energy/Energy Information Administration)

United States

As of 2008 the United States will export approximately 70 percent of its total coal production. The value of U.S. coal exports is approximately 3.75 billion dollars annually. The United States exports to more than 40 countries around the world. Currently, Canada, Japan, and Italy are among the biggest customers receiving shipments.

COAL IMPORTERS

Japan has continued to rely on coal and is expected to remain the world's largest coal importer. South Korea also is expected to continue importing most of the coal it consumes. With planned increases in coal-fired generating capacity, South Korea and Taiwan together are projected to maintain a roughly 16 percent share of world imports in 2030, despite sizable increases in coal imports by other countries. India's coal imports in 2030 are projected to be three times the 2007 level.

Italy's conversion of power plants from oil to coal also is projected to increase its coal imports, and Germany's planned closure of its remaining hard coal mines by 2018 is expected to result in increasing imports of coal for electricity generation. Israel also imports coal to meet approximately 25 percent of its energy requirements, primarily for electric power generation.

Top Coal Importers (2007)

	Total of which	Steam	Coking
Japan	182Mt	128Mt	54Mt
Korea	88Mt	65Mt	23Mt
Chinese Taipei	69Mt	61Mt	8Mt
India	54Mt	31Mt	23Mt
UK	50Mt	43Mt	7Mt
PR China	48Mt	42Mt	6Mt
Germany	46Mt	36Mt	10Mt

(*Source:* U.S. Department of Energy/Energy Information Administration)

COAL CONSUMPTION

The biggest market for coal is Asia, which currently accounts for 56 percent of global coal consumption, although China is responsible for a significant proportion of this. Many countries do not have natural energy resources sufficient to cover their energy needs and therefore need to import energy to help meet their requirements. Japan, Taiwan, and South Korea, for example, import significant quantities of steam coal for electricity generation and coking coal for steel production.

According to the World Coal Institute, about 36 percent of the world's electricity is produced by burning coal. Coal is a major fuel for generating electricity in Poland (97% of electricity), South Africa (93%), Australia (85%), China (80%), India (75%), and the United States (49%). In the year 2010, coal use was expected to rise in Southeast Asia, where coal was to be the major fuel for producing electricity.

As noted previously, most of the coal reserves are found in large deposits in the midlatitudes of the Northern Hemisphere, but there are a few coal deposits in the Southern Hemisphere. About 100 countries have coal reserves. Most of the world's largest deposits of coal are located in United States, Russia, China, Australia, and India. The largest producers and users of coal are China, the United States, India, and South Africa. Recent estimates indicate that the world's supply of coal should last for another 250 to 400 years, at current production levels.

ENVIRONMENTAL ISSUES

All coal-fired plants produce major air pollutants, such as carbon dioxide, sulfur dioxide, and nitrogen oxides, into the atmosphere. Studies indicate that about 70 percent of all sulfur dioxide emissions and 35 percent of carbon dioxide pumped into the atmosphere come from coal-burning power plants. Other air pollutants include volatile organic compounds (VOCs), soot, ash, and other particulate matter. Heavy metals such as cadmium and mercury are also released from coal-burning plants. They also produce bottom ash that needs to be collected and disposed of in landfills.

Sulfur Dioxide Emissions

Sulfur dioxide is a colorless gas with a characteristic acrid odor at high concentrations that is a common pollutant emitted when fossil fuels containing sulfur are burned.

Major emissions of sulfur dioxide in the United States derive from power plants east of the Mississippi River, particularly those in the Ohio Valley. When released into the atmosphere, sulfur dioxide reacts with water vapor to form sulfuric acid—a major component of acid rain. Sulfate particles also can be deposited as a dry contaminant that reacts with moisture in soil to form sulfuric acid. Acids formed from sulfur dioxide can be damaging to plants, aquatic ecosystems, and structures made from rock and metal.

Reducing Sulfur Dioxide Emissions

Today coal companies mine in areas where there is low-sulfur coal instead of high-sulfur coal when economically feasible. Using low-sulfur coals can reduce sulfur dioxide emissions. Washing coal to remove its sulfur is also effective but expensive. Some power plants use special equipment called scrubbers to remove most of the sulfur from coal emissions before they enter the atmosphere. One option that eliminates emissions of sulfur includes coal gasification, a process used by manufacturing plants to "scrub" the gas and remove sulfur compounds.

Carbon Dioxide Emissions

Environmental scientists are concerned because the amount of carbon dioxide (CO_2) in the atmosphere is increasing, and more is being released into the atmosphere than ever before. As global energy consumption increases, so will CO_2 emissions. According to some environmental reports, CO_2 emissions are projected to rise from 30 billion tons in 2006 to 36.1 billion tons in 2015 and 44 billion metric tons in 2030—an increase of 39 percent over this projection period if the data is correct.

Pulverized coal-fired burners dominate the power industry today, and a typical power plant can pump out more than six million tons of CO_2 a year—a major cause of global warming and the main greenhouse gas. And the addition of new power plants could bring an increase in future carbon dioxide emissions.

As a result of concerns about harmful emissions of coal burning, particularly the burning of high-sulfur and low-quality coal, the United States, Europe, and Japan initiated research and development programs in the 1980s to generate technologies, projects, and devices for controlling harmful emissions. These programs also sought ways to increase the efficiency of coal combustion, such as using clean coal technologies.

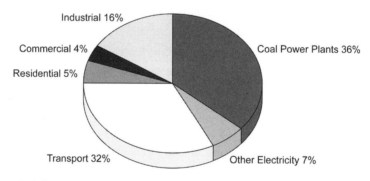

U.S. CO_2 Emissions by Sector and Source 2030

(*Source:* U.S. Department of Energy/Energy Information Administration *Annual Outlook 2009* projections)

 DID YOU KNOW?

Additional increases in carbon dioxide levels are caused by deforestation. When trees are cut down and burned to clear land for agriculture, CO_2 is released. In addition, as forests are cut down, there are fewer trees and other green plants to absorb carbon dioxide from the atmosphere.

CLEAN COAL TECHNOLOGY

In the United States, the Clean Coal Technology Program legislation was launched in 1986. Supervised by the U.S. Department of Energy, the program directed industry to burn coal more efficiently and to reduce emissions from domestic coal-fired plants. The program also sought ways to reduce the release of acid rain pollutants. One of the options in clean coal technology programs is coal gasification, the conversion of coal to a gaseous product by one of several available technologies; this is considered an alternative fuel.

Fluidized Bed Combustion Systems

The new Spurlock Fossil Plant in Maysville, Kentucky, uses clean coal technology known as the circulating fluidized bed combustion process. The new 268-megawatt plant is one of the cleanest coal-powered plants in the United States.

How Does the Fluidized Bed Combustion System Work?

Fluidized bed combustion is a low-polluting technology for burning low-grade coal in a boiler that traps sulfur dioxide emissions before they are emitted into the atmosphere. The technology was created through research and development sponsored by the Department of Energy.

According to the Department of Energy, the fluidized beds suspend solid fuels on upward-blowing jets of air during the combustion process. A mixture of pulverized coal and limestone is forced into the boiler, where it "floats" on the air while it burns. The calcium and some magnesium from the limestone absorb the sulfur dioxide from the sulfur materials in the coal. The result is a mixing of gas and solids. The tumbling action, much like a bubbling fluid, provides more effective chemical reactions and heat transfer.

Fluidized bed combustion evolved from efforts to find a combustion process able to control pollutant emissions without external emission controls such as scrubbers. The technology burns fuel at temperatures of 1,400

DID YOU KNOW?

America's coal-based electricity power companies have invested billions of dollars in technologies to reduce emissions. As a result, the coal-generating plants are much cleaner on the basis of regulated emissions per unit of energy produced.

to 1,700 degrees Fahrenheit, well below the temperature where nitrogen and oxygen atoms combine to form nitrogen oxide pollutants.

In summary, it is the mixing action of the fluidized bed that brings the flue gases into contact with a sulfur-absorbing chemical, such as limestone or dolomite. More than 95 percent of the sulfur pollutants in coal can be captured inside the boiler by the sulfur-absorbing chemical.

The popularity of fluidized bed combustion is due largely to the technology's fuel flexibility—almost any combustible material, from coal to municipal waste, can be burned—and the good news is that the process has the capability of meeting sulfur dioxide and nitrogen oxide emission standards without the need for expensive add-on emission control systems.

Other Coal-Cleaning Processes

As part of the Department of Energy's Clean Coal Technology Program, Custom Coals International of Pittsburgh, Pennsylvania, is testing coal-cleaning processes that will produce low-cost coals with low sulfur content. One such process crushes and screens mined coal and then applies a separation technique to remove about 90 percent of the sulfur. This special coal product can reduce sulfur dioxide emissions to levels that meet the compliance standards set by the new Clean Air Act.

Carbon Capture and Sequestration

Scientists are looking for new ways to use this old fuel and at the same time find ways to reduce, eliminate, or store the CO_2. One promising program is called "carbon capture and sequestration." Carbon sequestration is a plan for the long-term storage of carbon dioxide or other forms of carbon. It would take the CO_2 out of power plant emissions and store it deep in the earth.

One method to help reduce the CO_2 and store it underground is called the integrated gasification combined cycle (IGCC). There are only two IGCC plants now operating in the country—both gasify coal.

 **VIDEO**

Coal Comeback: A new generation of coal power stations promises zero emissions. To learn more watch the this video from Hitachi: http://videos.howstuffworks.com/hitachi/712-coal-comeback-video.htm (03:58 minutes).

DID YOU KNOW?

Twenty-first-century coal plants emit 40 percent less CO_2 than the average 20th-century coal plants.

Gasification begins when coal slurry is injected with oxygen into a high-pressure environment to create a gas. A turbine burns it to make power. Pollutants such as sulfur and particulates are stripped from the pressurized gas during the gasification. The same method could be used to strip out CO_2 as well. The leftover CO_2 could be buried in the ground or used in some places to force oil and gas from old wells. The carbon dioxide could also be pumped deep into formations in the earth where high pressure would make it a liquid. Small-scale tests show that the liquid CO_2 stays underground.

One study found that most parts of the country could store large amounts of CO_2 underground. Some of these places include aquifers, reservoirs, aging oil fields, or other carbon sinks. Now, the Department of Energy's main coal project is a billion-dollar partnership with coal and power companies called FutureGen. FutureGen would be the first IGCC plant with zero emissions.

 FEATURE

The National Energy Technology Laboratory

The National Energy Technology Laboratory (NETL) is part of the Department of Energy's national laboratory system, and it runs a project called the Carbon Sequestration Program. This program is helping to develop technologies to capture, separate, and store carbon dioxide (CO_2) in order to reduce greenhouse gas emissions without hindering economic growth. Carbon sequestration technologies capture and store CO_2 that would otherwise reside in the atmosphere for long periods of time.

Worldwide CO_2 emissions from human activity have increased from an insignificant level two centuries ago to annual emissions of more than 33 billion tons today. The U.S. Energy Information Administration predicts that, if no action is taken, the United States will emit approximately 7,550 million tons of CO_2 per year by 2030, increasing 2005 emission levels by more than 14 percent. The Carbon Sequestration Program contributes to President Barack Obama's goal of developing technologies to substantially reduce greenhouse gas emissions.

Clean Coal Technology in Germany

As mentioned, the concept of clean coal technology is to prevent carbon dioxide emissions from entering into the atmosphere from a power plant. Instead the carbon dioxide emissions are captured, collected, and then pumped deep into a natural rock formation for permanent storage.

In theory clean coal technology sounds good. However, such a system has not been built until recently.

In 2009 Germany built the first clean coal-fired power plant in the eastern region of the country. The pilot plant captures the carbon emissions before the gases are released into the atmosphere. First the coal is burned, producing boiling water to make steam that drives a turbine to generate electricity. The leftover carbon dioxide emissions and other materials are cycled back into a boiler. Under great pressure, the carbon dioxide is condensed into a liquid. The liquid is then transported to a closed natural gas site and pumped a few thousand feet deep into the depleted gas reservoir for storage.

Schwarze Pumpe is the world's first coal-fired plant ready to capture and store its own CO_2 emissions. The Schwarze Pumpe coal power plant is located in Spremberg, Germany. (AP Photo/Matthias Rietschel)

U.S. Funding for Sequestration Technology

The United States is also doing research on sequestration technology. In 2009 Energy Secretary Steven Chu announced that more than $8.4 million in funding had recently been approved for seven sequestration technology projects across the country and at some colleges and universities. "These projects will train workers for a clean energy economy and help position the United States as a leader in carbon capture and storage technologies for years to come," said Secretary Chu.

The National Energy Technology Laboratory of the Department of Energy (DOE) in the past has also offered numerous grants for carbon sequestration technologies and research. The national lab is responsible for advancing carbon storage research and projects. As recently as October 2009, it offered grants for Geological Sequestration Training and Research for up to $300,000. The Department of Energy also awarded more than $1 million for Wyoming professors to study carbon dioxide plume movements within storage reservoirs. Officials say the goal is to develop technologies for underground storage of the carbon dioxide that is emitted by power plants and other facilities that burn coal and gas.

Coal Gasification Technology: Synthetic Natural Gas

Coal gasification is the process of converting coal into a gas. The process results in what is known as synthetic natural gas. Synthetic natural gas fuel burns cleaner than coal and can be transported by pipeline. Gasification works by mixing coal with oxygen, air, or steam at very high temperatures to form methane, the major ingredient of natural gas. As a result, gasification plants allow for significant reductions in pollutants. For example, carbon dioxide emissions can be reduced by approximately 20 percent using coal gasification technology.

The Process

Step 1: Gasification. Gasification turns coal into a very hot, up to 3,200 degrees Fahrenheit, synthetic gas, or syngas, which is composed of carbon monoxide, hydrogen, and carbon dioxide, as well as small amounts of other gases and particles. This is accomplished by mixing pulverized coal with an oxidant, usually steam, air, or oxygen.

Step 2: Cooling and Cleaning. Next, the syngas is cooled and cleaned to remove the other gases and particles, leaving only carbon monoxide, carbon dioxide, and hydrogen. Syngas is easier to clean than the emissions from a pulverized coal power plant. During syngas cleaning, mercury, sulfur, trace contaminants, and particulate matter are removed.

Step 3: Shifting. Next, the syngas is sent to a "shift reactor." During the shift reaction, the carbon monoxide is converted into more hydrogen and carbon dioxide by mixing it with steam. Afterward, the syngas consists mostly of hydrogen and carbon dioxide.

Step 4: Purification. Once the syngas has been shifted, it is separated into streams of hydrogen and carbon dioxide. The hydrogen, once cleaned, is ready for use. The carbon dioxide is captured and sent off for sequestration.

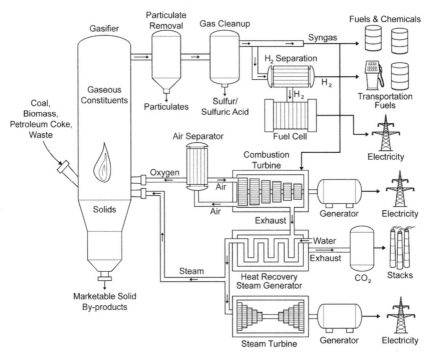

The heart of a gasification-based system is the gasifier. A gasifier converts hydrocarbon feedstock such as coal into gaseous components by applying heat under pressure in the presence of steam. The gasification power plant can also use feedstocks such as biomass and petroleum to produce synthetic gas. (Illustrator: Jeff Dixon)

Step 5: Usage. There is now a stream of pure hydrogen ready for a variety of uses. It can be burned in a gas turbine for electricity generation, converted to electricity in a fuel cell, used as a fuel for an internal combustion engine, or used as a chemical for making fertilizer, semiconductors, and many other valuable energy products.

Coal-gasification electric power plants are now operating commercially in the United States and in other nations, and many experts predict that coal gasification will be at the heart of future generations of clean coal technology plants.

Environmental Benefits

The environmental benefits of gasification are that the process can achieve extremely low sulfur dioxides, nitrogen dioxides, and emission particles from burning coal-derived gases. During gasification, sulfur in coal, for example, is converted to hydrogen sulfide and can be captured by processes presently used in the chemical industry. In some methods, the sulfur can be extracted in either a liquid or a solid form that can be sold commercially.

In an integrated gasification combined cycle (IGCC) plant, the syngas produced is virtually free of fuel-bound nitrogen. Nitrogen oxide from the gas turbine is limited to thermal nitrogen oxide. Diluting the syngas allows for nitrogen oxide emissions as low as 15 parts per million.

Using Synthetic Gas for Hydrogen Production: Hydrogen Fuel Cells

Gasification, in fact, may be one of the most flexible technologies for producing clean-burning hydrogen for tomorrow's automobiles and power-generating fuel cells for homes and businesses. Hydrogen and other coal gases can also be used to fuel power-generating turbines, or as the chemical "building blocks" for a wide range of commercial products.

 DID YOU KNOW?

Synthetic fuels, or synfuels, can also be made from oil shale and biomass (animal and plant wastes), but most are produced from solid coal.

Some energy companies are considering using gasification of coal for the production of hydrogen. The resulting syngas, when burned, produces nearly twice as much usable energy as coal.

Although syngas is a cleaner-burning energy source than coal, it does produce carbon dioxide, a major greenhouse gas associated with global warming. So one company plans to break down the syngas into its two components, hydrogen and carbon monoxide. The next step is to chemically change the carbon monoxide to make carbon dioxide. This process produces excellent sources of both hydrogen and carbon dioxide. The hydrogen is used to operate fuel cells. The carbon dioxide could be stored underground in a process called sequestration.

Not everyone believes that clean coal technologies are sufficient in reducing air pollutants. Many environmentalists believe the best way to reduce air pollutants is to replace coal-fired plants with those that use cleaner-burning fossil fuels or to eliminate the plants entirely and use renewable energy sources such as wind power and solar energy. But the problem is that several nations with large coal reserves want to build up their economies. These countries may find it cheaper to continue to use coal rather than switch to renewable technologies, which they might feel cost too much to use or take too long to develop.

THE FUTURE OF COAL

Coal will continue to play a key role in the world's energy mix, with demand in certain regions set to grow rapidly. Growth in the coal markets will be strongest in developing Asian countries, where demand for electricity and steel for construction and car production will increase as incomes rise.

Environmental Issues Still Prevail

Even though there have been major accomplishments in producing cleaner coal, some energy experts are predicting less coal use in the future because of emission regulations. As mentioned earlier, environmental scientists are concerned because the amount of carbon dioxide in the atmosphere is increasing; more is being released into the atmosphere than ever before.

Other energy experts favor the installation of new gas plants because even an efficient coal-fired power plant emits twice the carbon dioxide of a natural gas–fired plant. And there are a number of people who recommend eliminating coal-fired power plants altogether by 2030. Will coal be a dominant energy source for the next several decades? There is plenty of coal in the ground, and it may last more than 200 years at the present rate of consumption. But time will tell the story of coal as a major energy source after 2030.

BOOKS AND OTHER READING MATERIALS

Burns, Shirley Stewart. *Bringing Down the Mountains: The Impact of Mountaintop Removal on Southern West Virginia Communities.* Morgantown: West Virginia University Press, 2007.

Coal Age Magazine. http://coalage.com.

Freese, Barbara. *Coal: A Human History.* Cambridge, MA: Perseus Book Group, 2003.

Mahlum, D. D., et al., eds. *Coal Conversion and the Environment: Chemical, Biomedical, and Ecological Considerations.* Washington, D.C.: U.S. Department of Energy, 1981.

Riddle, John. *Coal Power of the Future.* New York: Rosen, 2003.

U.S. Geological Survey National Coal Resources Data System. http:energy.er.usgs.gov/coalqual.htm.

van Krevelen, D. W. *Coal: Typology—Physics—Chemistry—Constitution.* 3rd ed. Maryland Heights, MO: Elsevier Science.

SOMETHING TO DO

Synthetic fuels technology has been explained as a means to produce clean, reliable, and sustainable energy sources for the 21st century. The conversion of coal to liquid forms of energy is one of those technologies. There is serious debate as to whether liquid coal should be an option for reducing America's dependence on oil. Explore the arguments for and against the transformation of coal into diesel and other liquid fuels. Take a side and explain your decision. For some references, visit www.futurecoalfuels.org and www.nrdc.org/globalwarming/solutions.

WEB SITES

The following Web sites, though not inclusive, include government and non-government organizations.

www.americancoalcouncil.org
   The American Coal Council (ACC) is dedicated to advancing the development and utilization of American coal as an economic, abundant/secure, and environmentally sound energy fuel source.

www.coaleducation.org/miningtv/modern_videos.htm
> Kentucky's coal education Web site presents factual, useful information about coal in a fun and productive way.

web.mit.edu/coal/.
> The Massachusetts Institute of Technology (MIT) faculty group examines the role of coal in a world where constraints on carbon dioxide emissions are adopted to mitigate global climate change.

http://fossil.energy.gov/
> The primary mission of the Department of Energy's Office of Fossil Energy is to ensure that we can continue to rely on clean, affordable energy from our traditional fuel resources.

VIDEOS

The following video and audio selections are suggested to enhance your understanding of coal energy topics and issues. The author has made a consistent effort to include up-to-date Web sites. However, over time, some may move or may no longer be available.

Viewing some of these videos may require special software called plug-ins. Therefore, you may need to download that software to view the videos. You also may need to upgrade your player to the most current version.

> **What is coal and how is electricity produced?** To learn more about the earth's original biofuel, go to http://www.schooltube.com/video/39882/Peabody-Energy-What-Is-Coal (02:33 minutes).
>
> **Coal Energy—Duke Power, United States:** Caught between a rock and a hot place! The CEO of Duke Power, America's third-largest energy provider, discusses the daunting challenges ahead for the coal industry to lower CO_2 emissions. Clean coal makes putting a man on the moon look easy. To learn more about how the United States could adapt its 200-year reserve to benefit the globe and preserve the American way of life, go to http://www.cbsnews.com/video/watch/?id=4969902n (10:00 minutes).
>
> **Coal Comeback**—A new generation of coal power stations promises zero emissions. To learn more, watch the following video from Hitachi: http://videos.howstuffworks.com/hitachi/712-coal-comeback-video.htm (03:58 minutes).

Chapter 5

Nuclear Energy

In 2009 U.S. Energy Secretary Steven Chu announced the selection of 71 university research project awards. The awards were part of the Department of Energy's investments in nuclear energy research and development.

Under the Nuclear Energy University Program (NEUP), the cost of these projects will be about approximately $44 million when completed. The NEUP program will help advance new nuclear technologies in support of the nation's energy goals and will play a key role in addressing the global climate crisis and moving the nation toward greater use of nuclear energy.

"As a zero-carbon energy source, nuclear power must be part of our energy mix as we work toward energy independence and meeting the challenge of global warming," said Secretary Chu. "The next generation of nuclear power plants—with the highest standards of safety, efficiency and environmental protection—will require the latest advancements in nuclear science and technology. These research and development university awards will ensure that the United States continues to lead the world in the nuclear field for years to come."

In 2010 President Barack Obama announced an $8.3 billion federal loan to build two new reactors in Georgia. "We'll have to build a new generation of safe, clean nuclear power plants in America," said President

FEATURE

The U.S. Department of Energy

The Department of Energy's nuclear energy program is designed to promote secure, competitive, and environmentally responsible nuclear technologies to serve the present and future energy needs of the United States and the world.

Obama. This was significant news given that there have been no new nuclear units licensed since the near-meltdown at the Three Mile Island nuclear power plant near Middletown, Pennsylvania, in 1979.

WHY THE INTEREST IN NUCLEAR ENERGY?

One answer to this question is that the global demand for electricity is expected to increase by almost 50 percent by 2030, according to the U.S. Department of Energy. As of 2009, nuclear energy provides almost 20 percent of all electricity used in the United States and is responsible for about 15 percent of the world's electrical energy output, according to the World Nuclear Industry Report. However, the construction of new nuclear power plants in the United States could provide 33 percent of U.S. electricity, according to advocates for nuclear power.

The current conventional sources of electric power, such as coal, natural gas, and hydropower, may not be able to supply all of the world's electrical needs by 2030. Additionally, the renewable energy sources such as wind, solar, and geothermal may still lag behind as major sources of electricity during this time. In fact, presently, the renewable non-hydropower fuels supply less than 3 percent of electrical energy needs in the United States, according to the U.S. Energy Information Administration.

One of the major benefits of nuclear energy is that nuclear power plants can operate without contributing to climate change. Although the complete nuclear fuel cycle emits small amounts of greenhouse gases because of the fossil fuels used to mine uranium, transport nuclear fuel, and provide some of the electrical energy to run uranium enrichment plants, the ratio of greenhouse gases emitted to the electricity generated is lower for nuclear energy than for virtually all other electricity generation sources. (The possible exceptions may be certain hydropower and geothermal

The dome of the nuclear reactor of Sizewell nuclear power plant in eastern England, commissioned in 1995. According to the World Nuclear Association, the United Kingdom has 19 reactors generating about 15 percent of its electricity; all but one of these will be retired by 2023. (iStockphoto)

plants, but these sources confront geographical and environmental limitations.) When operating, nuclear power plants do not emit greenhouse gases. However, there are safety, security, and environmental issues related to using nuclear power as a major energy supplier in the future. Nonetheless, because of nuclear power's benefits, several countries are now initiating new and extensive nuclear energy projects and programs to help supply more worldwide electricity demands in the future.

WHAT IS NUCLEAR ENERGY?

Nuclear energy is the energy stored within the nuclei of atoms. Atoms are tiny particles that make up every object in the universe. There is enormous energy in the bonds that hold atoms together. However, the energy is very concentrated, so it is hard to release it unless a nuclear reactor is used.

Inside the nuclear reactor, the nuclei of heavy atoms of an element, such as uranium or plutonium, are split apart to form smaller atoms, releasing

 DID YOU KNOW?

Nuclear energy is used to detect and treat certain illnesses.

energy when struck by a neutron. The splitting or "the visioning" of the nucleus releases more atoms and energy, both at the same time. When this action occurs, energy is released. The neutrons continue to split other nuclei to cause a chain reaction, a series of nuclear fissions that produce enough neutrons to keep the reaction going.

The heat energy given off during fission in the reactor is used to boil water into steam, which turns the turbine blades. As they turn, they drive generators that make electricity. Afterward, the steam is cooled back into water in a separate structure at the power plant called a cooling tower. The water can be used again and again.

The electricity is transmitted along transmission power lines carried to communities. The U.S. Nuclear Regulatory Commission (NRC) regulates the safe and secure operation of all U.S. nuclear power plants.

HISTORY OF USING NUCLEAR ENERGY

The first experimental fission reaction was observed in December 1938 by chemists Otto Hahn and Fritz Strassman in Germany, but the results were not fully understood until January 1939, when physicists Otto Frisch and Lise Meitner explained the reaction as fissioning of uranium nuclei. In 1942 at the University of Chicago, scientists, led by Enrico Fermi, produced a self-sustaining chain reaction using nuclear fuel.

During World War II, the U.S. government approved "The Manhattan Project," a top-secret program conducted between 1942 and 1945 for the purpose of developing an atomic (nuclear) bomb for use during World War II. As a result of the Manhattan Project, three atomic bombs (Trinity Test, Fat Man, and Little Boy) were produced. In the summer of 1945, two atomic bombs were dropped on the cities of Hiroshima and Nagasaki, Japan.

After World War II, a major effort was made to apply nuclear energy to nonweapon uses. The world's first commercial-scale nuclear reactor power plant began to operate in Britain in 1956. In 1957 in Shippingport, Pennsylvania, the first large-scale nuclear power plant, a pressurized water reactor, began operations.

The U.S. program expanded quickly in the 1960s and 1970s. In 2009, 104 power reactors supplied about 800 billion kilowatt-hours of electricity in the United States, almost 20 percent of total electricity for the country.

The Shippingport Atomic Power Station in Shippingport, Pennsylvania, was the country's first large-scale civilian atomic power plant to generate electricity for commercial use. (Library of Congress)

CONCERNS ABOUT NUCLEAR WEAPONS

During the 1960s and 1970s, many nations were concerned about the buildup of nuclear weapons and the proliferation of these weapons to other nations. The concern centered on the misuse of nuclear energy—a dual-use technology—for weaponry. In 1970 the Treaty on the Nonproliferation of Nuclear Weapons went into effect. Nations without nuclear weapons agreed not to develop them in exchange for the provision of non-nuclear materials and technology from the nations that already had nuclear weapons. In a major effort to limit the nuclear arms race between the United States and the Soviet Union, negotiations such as the Strategic Arms Limitation Talks (SALT) were pursued during the 1980s.

INTERNATIONAL ATOMIC ENERGY AGENCY

The International Atomic Energy Agency (IAEA) attempts to ensure that countries do not misuse peaceful nuclear programs to make weapons. It is a United Nations agency, headquartered in Vienna, Austria, that promotes safe and secure use of peaceful nuclear energy. The IAEA advises countries on such matters as nuclear energy, nuclear radioactive waste management, and nuclear safety and security programs.

WORLD USE OF NUCLEAR ENERGY

On the global scene, as of 2009, 31 countries, including the United States, have chosen nuclear power as part of their energy needs. Other countries that use nuclear energy include China, Russia, France, Belgium, Germany, India, Japan, Poland, and South Korea. According to the World Nuclear Association, there were 444 nuclear power reactors worldwide in 2009. These reactors supply approximately 17 percent of the world's electrical needs for more than one billion people without emitting any carbon dioxide or other greenhouse gases during their operation.

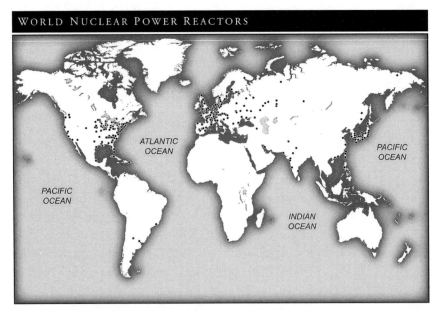

Many of the world's nuclear power reactors are located in the United States and in Europe. In recent years France has emerged as a leader in the nuclear industry. (*Source:* International Nuclear Safety Center at Argonne National Laboratory, 2005)

United States

The United States is still the largest single producer of nuclear energy in the world, with 104 units supplying more than 750 billion kilowatt (kW) hours as of 2009. The United States is using more nuclear energy today as a result of improving equipment, procedures, and general efficiency, without a new reactor order. (Watts Bar Unit 1, completed in 1996, was the latest completed U.S. reactor as of 2010.)

There are currently 31 states with nuclear power plants. Some of these states include Washington, Texas, Tennessee, Illinois, Florida, and California. Most plants go offline for relatively short periods of time—less than one month annually—for refueling and basic maintenance. However, some plants may go offline for extended periods to make more substantial repairs.

Of the 31 states, 6 rely on nuclear power for more than 50 percent of their electricity, and another 13 states rely on nuclear power for up to 25–50 percent of their electricity. However, determining which states use nuclear power can be difficult. Because power generation is shared across state lines, it is reasonable to assume a vast majority of states, if not all, use nuclear power to some extent.

France

According to the Nuclear Energy Agency, as of 2009, France had the second-largest number of commercial reactors with 59, and it was building one new reactor at Flamanville, with plans for another new reactor at Penly. France is a major global producer of nuclear power for electricity. France's first nuclear reactor began operating in 1974; the most recent one (before Flamanville and Penly) in 2000. About 78 percent of France's electricity is produced by nuclear energy. France is a major exporter of electricity to other countries in Europe.

  DID YOU KNOW?

In addition to state nuclear power plants, the U.S. military is a big user of nuclear power. Submarines and naval ships often have nuclear power plants as their primary power source. This is particularly true for newer vessels.

>
>
> **France's Nuclear Energy Program.** With nuclear power producing nearly 80 percent of the country's energy needs, France has more than 55 nuclear reactors in a country the size of Texas. To learn more, go to http://www.youtube.com/watch?v=i-rKBrs7kYE&feature=related (10:00 minutes).

Other Countries

As of 2009, South Korea had 20 reactors providing almost 40 percent of the country's electricity. Japan's 53 reactors produce about 30 percent of the country's electrical needs. Russia uses 31 reactors for generating about 16 percent of its electricity. Across Europe there are about 192 reactors in such countries as Belgium, Germany, Lithuania, and Poland. Today, additional nuclear power plants are under consideration, in the planning stages, or under construction in Russia, China, and India, to name a few.

NUCLEAR FUEL

Nuclear plants use uranium fuel, consisting of solid ceramic pellets. The basic fuel used in nuclear reactors is uranium-235 (U-235). In nature, less than 1 percent of uranium is in the form of isotope U-235. It occurs mixed in with uranium's much more abundant form, U-238, which constitutes more than 99 percent of natural uranium. For most commercial reactors, the concentration of U-235 has to be increased through an industrial process called enrichment in order to be able to sustain a fission chain reaction. Following is an example of this fission reaction:

$$_{92}U^{235} + {}_0n^1 \rightarrow {}_{36}Kr^{90} + {}_{56}Ba^{142} + \text{neutrons}$$

This chemical reaction in the fuel releases several neutrons per atom of U-235, making a chain reaction possible. (Kr-90 and Ba-142 are two examples of fission products; other common fission products include Cs-137 and Sr-90.) In a nuclear reactor, the fuel's chain reaction is controlled to maintain a steady reaction rate.

Mining Uranium

Uranium minerals are widely distributed in Earth's crust. They are present in sandstones, in veins within rock fractures, and in placer deposits—ore

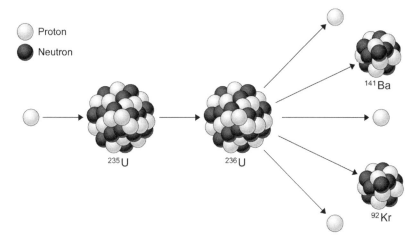

Nuclear energy is energy in the nucleus (core) of an atom. In nuclear fission, atoms are split apart to form smaller atoms, releasing energy. Nuclear power plants use this energy to produce electricity. (Illustrator: Jeff Dixon)

materials that have been transported and deposited in river deltas and streams.

Most uranium mined in the United States derives from sandstone deposits. Worldwide, the richest deposits of uranium are in Russia, Ukraine, Australia, Canada, and southern Africa. According to the Nuclear Energy Agency, known uranium deposits could fuel the current global reactor fleet for at least another 80 years. The total amount of uranium is unknown. As the price of uranium goes up, there is more incentive for additional prospecting to find more deposits. The world's oceans could conceivably supply several hundred years or more of uranium based on current demands for nuclear-generated electricity.

According to the Department of Energy, in 2008 the major foreign suppliers of uranium to the United States were Australia and Canada (42%), followed by Kazakhstan, Russia, and Uzbekistan (33%) and finally, Brazil,

 DID YOU KNOW?

A single, quarter-ounce pellet of uranium creates as much energy as three and a half barrels of oil or 1,780 pounds of coal, without carbon dioxide emissions.

Czech Republic, Namibia, Niger, South Africa, and the United Kingdom (11%). Only 14 percent of the delivered uranium to nuclear power plants was from uranium ore deposits in the United States. Two of the states with the largest known uranium ore reserves include Wyoming and New Mexico.

Processing Uranium to Make Fuel

The major nuclear reactors use essentially the same uranium fuel. However, before uranium can be used in a reactor, it must undergo several processing steps to be converted from an ore to solid ceramic fuel pellets. The pellets are about the size of a human's fingertip, but each one produces roughly the same amount of energy as 150 gallons of oil.

The processing steps include mining and milling, conversion, enrichment, and fabrication. Uranium miners obtain uranium ore via surface or open-pit mining and underground mining. Special liquid cleaners remove the uranium from the mined ore, and the resulting uranium oxide is called yellowcake.

The yellowcake is filtered and dried and then is converted in a chemical processes to uranium hexafluoride. Uranium hexafluoride is a chemical compound of uranium consisting of one atom of uranium combined with six atoms of fluorine. The uranium hexafluoride is processed as a gas. The enriched uranium hexafluoride is stored in cylinders. When it cools, it condenses into a solid. Through another chemical process, the uranium hexafluoride is transformed into uranium dioxide, which is used to make fuel rods.

Fuel Rods

The fuel used in most nuclear reactors is natural uranium oxide or enriched uranium oxide U-235. (Note: Some reactors are fueled with mixed oxide fuel, which combines uranium oxide and plutonium oxide.) U-235 is one of the fissionable isotopes of uranium. The enriched fuel is made into the ceramic pellets and placed inside fuel rods made of a zirconium alloy, or other material. The fuel rods are joined together in a reactor core. When the U-235 is bombarded with neutrons, fission reaction takes place in the reactor core:

$$_{92}U^{235} + {_0}n^1 \rightarrow {_{36}}Kr^{90} + {_{56}}Ba^{142} + \text{neutrons}$$

Uranium pellets fill fuel rods in this full-scale model of a nuclear fuel assembly at the Ulba Metallurgical Plant in Ust-Kamenogorsk, Kazakhstan. These energy-rich pellets are stacked end-to-end into long metal fuel rods. A bundle of fuel rods is called a fuel assembly. A reactor core contains many fuel assemblies. (Daniel Acker/Bloomberg via Getty Images)

In U-235 atoms, the nucleus is unstable. As the atoms' nuclei break up, they release neutrons. When the neutrons hit other uranium atoms, those atoms also split, releasing neutrons along with heat energy. These neutrons strike other atoms, splitting them, and they in turn split other atoms, until there is a chain reaction. When that happens, fission becomes self-sustaining.

Moderator and Coolants

The moderator in the nuclear reactor is used to slow down the neutrons. The right speed is maintained for a steady fission rate. The moderators contain a variety of materials, including pure water, heavy water or deuterium oxide, and graphite. Coolants are piped into and out of the reactor core, removing excessive amounts of heat that build up in the reactor. The steam generator is part of the cooling system in which the heat from the reactor is used to make steam for the turbine.

The list of coolants includes pure water or heavy water, carbon dioxide, sodium, and helium. Most commercial reactors use water as both coolant and moderator. The discharges of the heated water are pumped into cooling towers or nearby waterways.

Control Rods

The chain reaction is regulated by control rods made from neutron-absorbing materials such as cadmium or boron. The control rods in the reactor core are raised or lowered to speed up, slow down, or stop the fission. No combustion occurs in the process of producing nuclear energy, and therefore, no greenhouse gas emissions are released into the atmosphere.

THE KINDS OF NUCLEAR REACTORS

Commercial nuclear power plants in the United States are either boiling water reactors or pressurized water reactors. Both boiling water reactors and pressurized water reactors are cooled by ordinary water. The water is the main that carries the heat from the fission reaction to the generator that produces electricity. The other kind of reactor is the breeder reactor. The United States does not use breeder reactors.

Boiling Water Reactors

Boiling water reactors use fission to boil water and produce steam. The steam is transferred by pipes directly to the turbine, which drives the electric generator to produce electricity. The boiling water reactor obtains the water it needs from several sources, including rivers, lakes, streams, and oceans. The radioactive water flows back to the reactor core, where it is reheated and returned back to the steam generator. Approximately 30 of every 100 nuclear reactors in the United States are boiling water reactors.

 **VIDEO**

To view an animated boiling water reactor in action, visit http://www.nrc.gov/reading-rm/basic-ref/students/animated-bwr.html.

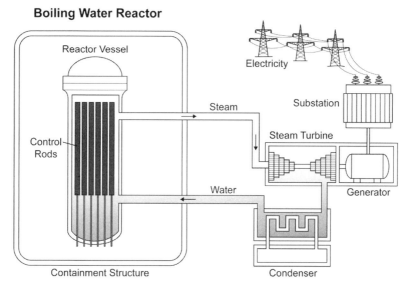

A boiling water reactor is a light water reactor in which water is used as both coolant and moderator and is allowed to boil in the core. The resulting steam can be used directly to drive a turbine to produce electricity. (Illustrator: Jeff Dixon)

Pressurized Water Reactors

Approximately 70 of every 104 reactors in the United States are pressurized water reactors. Pressurized water reactors heat water, similar to the other kinds of reactors. However, the pressurized water reactors keep the water under high pressure to prevent it from boiling. The hot water is pumped from the reactor to a steam generator. There, the heat from the water is transferred to a second, separate supply of water that boils to make steam. The steam spins the turbine, which drives the electric generator to produce electricity.

Breeder Reactor

The breeder reactor is quite different from the other reactors. A breeder nuclear reactor generates at least as much fissionable fuel as it consumes, while also producing steam that can drive a turbine and generator to produce electricity. Because they produce fuel, breeder reactors could greatly extend the useful life of uranium reserves.

Pressurized Water Reactor

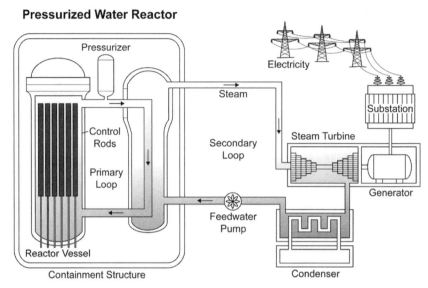

A pressurized water reactor is a nuclear reactor in which heat is transferred from the core to a heat exchanger by water. The water is kept under high pressure so that high temperatures can be maintained in the system without boiling the water. Steam is generated in a secondary circuit. (Illustrator: Jeff Dixon)

According to the Department of Energy, a breeder reactor is designed to produce fissionable plutonium-239 (Pu-239) from uranium-238 (U-238). The core of the reactor consists of bundles of plutonium-filled fuel rods surrounded by an outer layer of U-238 fuel rods. The U-238 fuel rods are bombarded by high-speed neutrons, which split apart the U-238 and cause a chain reaction in which some U-238 is transformed into Pu-239. This process can produce plutonium in sufficient amounts for extraction and processing for later use as fuel.

However, a disadvantage of breeder reactors is that the electricity they generate is generally more costly than that produced by other nuclear reactors. Breeder reactors are also expensive to construct and require a great deal more engineering to be made operational. Another disadvantage is that plutonium is much more radioactive than uranium, making disposal of breeder reactor wastes difficult and the threat of a nuclear disaster more imposing. Plutonium can also be used to build nuclear weapons.

SOME MAJOR NUCLEAR REACTOR MANUFACTURERS

Globally, General Electric and the Westinghouse Electric Company are some of the major manufacturers and suppliers of nuclear power plant products and technologies. In 1957, Westinghouse supplied the world's first pressurized water reactor in Shippingport, Pennsylvania.

General Electric designs the boiling water reactor that is used throughout the world. Its newly designed boiling water reactor will be installed in Japan. Other manufacturers developing newer designs of nuclear reactors are GE-Hitachi Nuclear Energy Canada, Areva Company in France, and Mitsubishi in Japan.

NUCLEAR POWER PLANTS GENERATE WASTE

Like all industrial processes, nuclear power generation has by-product wastes such as spent fuels and other radioactive wastes, which are the principal environmental concerns for nuclear power.

A nuclear plant can produce up to several hundred pounds of high-level radioactive wastes each year. Radioactive wastes can fall into several categories—high-level radioactive wastes, low-level radioactive waste, transuranic wastes, and uranium mill tailings.

High-Level Radioactive Wastes

Highly radioactive material waste associated with the production of weapons and in the operation of nuclear reactors is called high-level radioactive waste. This waste is no longer useful in sustaining a nuclear reaction.

One form of high-level radioactive waste is spent nuclear fuel. This is a solid material that is composed of uranium oxide pellets encased in metal tubes called fuel rods. This material is highly radioactive and very dangerous and must be stored safely. Thousands of tons of spent nuclear fuel are stored at commercial nuclear power reactors, and the amount grows each year. In fact, the average-sized nuclear power plant reactor produces 25 to 30 tons of spent fuel per year.

Low-Level Radioactive Wastes

However, according to the Department of Energy, "most nuclear waste is low-level radioactive waste. Uranium enrichment processes, contaminated lab equipment, isotope production, and research and development activities generate such waste."

Mostly, low-level wastes (LLW) include materials such as tools, clothing, rags, papers, filters, equipment, soil, and construction rubble that are contaminated with low levels of radioactivity. This waste does not include high-level wastes, transuranic waste, or uranium mill tailings.

Low-level wastes are less hazardous than high-level radioactive wastes because the radioactivity of most LLW diminishes to harmless levels through radioactive decay after several years. They require little or no shielding and no cooling during handling and transporting.

However, a small fraction of LLW are longer-lived radioactive substances that can pose public health risks for up to a few hundred years. Therefore, these materials are subject to special regulation that governs their disposal so that they will not come in contact with the outside environment.

Transuranic Wastes

Transuranic (TRU) waste materials have been generated in the United States since the 1940s. Most of this waste originates from nuclear weapons production facilities for defense programs. "Transuranic" refers to the "heaviness" of the element—these elements are heavier, meaning higher in atomic number in the periodic table, than uranium. Some of these heavy elements include plutonium, neptunium, americium, and curium. The most major element in most TRU waste is plutonium.

Some TRU waste consists of items such as rags, tools, glassware, protective clothing, and laboratory equipment contaminated with radioactive materials. Other forms of TRU waste include organic and inorganic residues or even entire enclosed contaminated cases in which radioactive materials were handled. These wastes decay slowly and need long-term waste storage.

Most of the TRU waste can be packaged and stored in metal drums or in metal boxes. They can be handled under controlled conditions without any shielding beyond the container itself. The waste emits primarily alpha particles that are easily shielded. However, about 3 percent of TRU waste must be both handled and transported in shielded casks. This TRU waste emits gamma radiation, which is very penetrating and requires concrete, lead, or steel to block the radiation.

Other radioactive wastes include uranium mill tailings from the mining and processing of uranium ore. The tailings consist of rock and soil containing small amounts of radium and other radioactive materials. Uranium mill tailings become a radioactive waste disposal problem because radon, a radioactive gas, is produced when radium decays.

U.S. Nuclear Regulatory Commission

The Nuclear Regulatory Commission is an agency established by the U.S. Congress under the Energy Reorganization Act of 1974. According to the agency's objectives, the NRC ensures protection of the public's health and safety and the environment in the use of nuclear materials in the United States. Other responsibilities include the regulation of commercial nuclear reactors, industrial uses of nuclear materials, and the transport, storage, and disposal of nuclear materials and waste.

Either the NRC or the States must license low-level waste disposal facilities in accordance with health and safety requirements. The facilities are to be designed, constructed, and operated to meet the safety standards required by the license. Those responsible for operating the facilities must analyze how the facility will perform for thousands of years into the future. As of 2009, the United States is facing a shortage of low-level waste disposal facilities.

Disposing of Nuclear Waste

Safely disposing of radioactive wastes that may remain radioactive for thousands of years is a major concern and top-priority issue for many countries that use nuclear energy. According to the IAEA, about 350,000 cubic feet of high-level waste accumulate each year. Unfortunately, not one country has any long-term plan or program for where to store radioactive wastes

DID YOU KNOW?

If the waste is transuranic and has a concentration of more than 100 nanocuries per gram, it is treated as transuranic waste. If the concentration is less than 100 nanocuries per gram, it is treated as low-level waste.

safely. The United States had investigated the possibility of constructing a high-level waste depository in Yucca Mountain in Nevada. However, the Obama administration decided not to fund the project in 2009; nevertheless, many members of Congress still favor using the facility. At that time, President Obama and Senate Majority Leader Harry Reid agreed to form a blue ribbon senior advisory panel to study other high-level waste disposal options. Both President Obama and Senator Reid have said that Yucca Mountain is not a viable option. Meanwhile, the NRC is continuing to review the license application for Yucca Mountain because as of 2009, U.S. law requires ongoing consideration of the project. Depending on the deliberations of the blue ribbon panel, the president and Congress may formally shut down the project, decide to continue it by itself, or decide to pursue another disposal facility with or without the Yucca Mountain option.

Nuclear Waste Policy Act

The U.S. government is responsible for finding a way to safely dispose of this spent nuclear fuel. In 1982 the U.S. Congress passed the Nuclear

A worker conducts an underground train in 2006 into a tunnel leading to the proposed Yucca Mountain nuclear waste dump near Mercury, Nevada. (AP Photo/Isaac Brekken)

Waste Policy Act, which directed the Department of Energy to find a site. Although sites in Texas and the state of Washington were considered, Yucca Mountain, in the southwestern part of Nevada, was selected in 1987.

More than two decades of work went into making a Yucca Mountain nuclear facility that would be used as a long-term repository for the nation's commercial and defense spent fuel and high-level radioactive wastes. The project is under the supervision of the U.S. Department of Energy, which, along with the U.S. Geological Survey, conducted studies to evaluate earthquake activity, ways to ensure habitat preservation, and the socioeconomic impact of this potential site. But several environmentalists and organizations voiced their objections to the Yucca program because they believe the stored radioactive wastes would be a health risk and a danger to future populations in the area, and as noted previously, in 2009 the Obama administration did not request funding for further construction of the project, leaving its continuation very much in doubt. In the future, a blue ribbon panel will assess high-level waste disposal options and report to the president and Congress.

Recycle Nuclear Waste

Can any of the spent fuel be recycled? One of the misconceptions of nuclear power is that once the fuel is removed from the commercial nuclear reactor, it has been used up. It turns out that there is still 95 percent usable energy in the spent nuclear fuel after it is removed from the light-water nuclear reactor used today. This spent fuel can be recycled and used in fast neutron reactors or used in a new generation of nuclear reactors to produce more energy.

Fast Neutron Reactors or Fast Breeders

According to nuclear experts, the development of fast neutron reactors can result in the creation of more energy and less nuclear wastes in comparison with the present conventional nuclear reactors. Fast breeders can also burn the recycled spent fuel wastes from other nuclear reactors. Therefore, this kind of reactor is more efficient in using nuclear fuel than other types of reactors. How is it done? Simply stated, fast neutron reactors extract more energy from nuclear fuel than other reactors because their fast-moving, higher-energy neutrons cause atomic fissions more efficiently than the slow-moving neutrons used in the other reactors. And in the process, a bit more usable new fuel, plutonium, is also created for use in the reactor.

Several countries have or are considering building fast neutron reactors. In fact, France is planning for half of its present nuclear capacity to be replaced by fast neutron reactors by 2050. China, Japan, and Russia are also planning on more advanced fast neutron nuclear reactors.

Other Plans to Transform Wastes into Nuclear Fuel

In the United States, scientists at the Oak Ridge National Laboratory in Tennessee have worked on plans to transform nuclear leftover wastes into fuel for a new breed of reactors. The new reactor and fuel, according to the scientists, could produce up to 100 times as much energy as a conventional reactor and generate 40 percent less waste than the conventional reactor.

Presently, countries such as France, the United Kingdom, Japan, and Russia use a process called plutonium uranium recovery by extraction

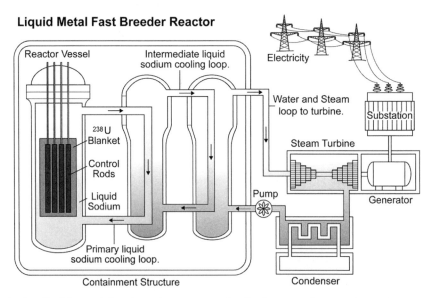

In the liquid metal fast breeder reactor, the fission reaction produces heat to run the turbine while at the same time breeding plutonium fuel for the reactor. (Illustrator: Jeff Dixon)

(PUREX) to recycle nuclear waste. The United States stopped recycling waste to support the nonproliferation policy because the PUREX process can be used to create weapons-grade plutonium.

Many of these technologies to recycle spent fuel are still in the research phase. The benefits of these technologies would create less waste while binding the plutonium to other chemicals so that it is much less desirable for use in weapons.

CUTTING-EDGE NUCLEAR TECHNOLOGIES

Pebble Bed Modular Reactor

Many companies are developing new technologies in building nuclear reactors. One such technology is called a pebble bed modular reactor (PBMR).

The PBMR is a steel pressure vessel that holds about 450,000 fuel spheres. The spheres consist of low-enriched uranium particles that are encased in graphite and are about the size of a baseball. The coated layers on the spheres provide a barrier that is dense enough to ensure that no radioactive products escape. The heat provides energy to the turbine's generators to produce electricity. The heat that is transferred by the helium to the power conversion system is converted into electricity through a turbine. The PBMR system is cooled with helium. More research is being conducted on this technology.

Floating Nuclear Power Plants

There are more than 200 nuclear reactors powering various kinds of ships. Russia has designed a plan to build a nuclear power plant on a large barge. The barge would be permanently moored, and the nuclear power plant would supply electricity to local shore communities and coastal industrial

 **VIDEO**

Nuclear waste and recycling: For more on nuclear waste recycling, refer to the following video, "Waste and Recycling," featuring nuclear engineer Dr. Kathy McCarthy: www.ne.doe.gov/video/videoGallery.html (2:20 minutes).

FEATURE

"Neutropolis: The Nuclear Energy Zone for Students"

The U.S. Department of Energy's Office of Nuclear Energy is extending its outreach efforts for youth with the launch of a new Web site geared toward students in grades K–12. The new site features a mythical city known as Neutropolis and was tested by select target audiences prior to posting.

"Neutropolis: The Nuclear Energy Zone for Students" provides information about nuclear energy and its many uses in separate tracks directed at younger, intermediate, and advanced students. It also contains games and challenges designed to appeal to students of different ages, ideas for science projects, information about colleges that offer studies related to nuclear science, and details about Department of Energy internships and career opportunities.

Although the primary audience for the site is K–12 students, there is a section devoted to providing information to classroom teachers, including suggested classroom activities targeted to elementary, middle, and high school students.

The Department of Energy hopes the new Web site "will set a new bar for quality outreach to students, delivering the message that nuclear energy is an important current and future part of our national's energy mix." For more information see http://www.nuclear.energy.gov/students/intro.html.

complexes in remote areas of the Russian Far North and East. For more information about this plan, see http://cns.miis.edu/stories/020624.htm.

Is There a Future for Nuclear Energy?

Most energy analysts would probably agree that a rapid growth of nuclear energy power plants in the next 20 years looks a bit challenging. Throughout the world, there is public opposition to new nuclear plants, concerns about safety standards and regulations, and the potential for accidents in the present plants. Other concerns include how to remove and dispose of existing nuclear waste as well as the challenge of decommissioning obsolete plants.

In 2008 the Nuclear Energy Institute (NEI) offered its annual outlook for the future of nuclear power, and it was optimistic. Today's 104 nuclear power plants in the United States generate about 20 percent of the country's electricity. Because of rising energy demands and aging infrastructure, the NRC predicts that industry will need to build 50 new reactors to continue producing the same proportion of the country's power over the next 30 years.

Thirty-one reactors, representing 17 power companies and consortia, are somewhere in the application process. The NEI predicts that only four to eight of those will be in commercial operation by 2016. By that time, pressure for an affordable, clean source of energy could inspire a second wave of applications.

However, other countries are more aggressive in viewing nuclear energy as another clean energy alternative to fossil fuels. According to the IAEA, there were several nuclear reactors under construction in 2009; additionally, India was planning to build eight reactors, China and Ukraine were set to build four each, Japan and the Russian Federation will build three each, Iran and South Korea will build two each, and Argentina and Romania are planning to construct one each.

The global demand for electricity is growing rapidly. If this trend continues, nuclear exponents may resurrect more creative technologies for developing new kinds of nuclear reactors and power plants to meet the growing needs for electricity.

Designing New and Different Kinds of Nuclear Reactors

There is a growing international consensus that to be broadly acceptable for the 21st century and beyond, the next-generation advanced reactor system must meet these five criteria:

- provide a long-term energy source not limited by resources
- be passively safe, based on characteristics inherent in the reactor design and materials
- reduce the volume and toxicity of nuclear waste
- keep nuclear materials unsuitable for direct use in weapons
- be economically competitive with other electricity sources

One type of reactor that can meet all five requirements simultaneously is the advanced fast reactor mentioned earlier. These fast reactors can extract more energy from the nuclear fuel than other kinds of nuclear reactors and can recycle spent fuel from other nuclear reactors. The advanced fast reactor produces extremely hot and high-pressured steam that is used to turn a steam turbine, which in turn drives generators to produce electricity. This kind of reactor produces clean, safe nuclear power and less waste and is unsuitable for use in weapons, according to the Department of Energy.

This new nuclear power technology may still be developed as the supply of fossil fuel resources dwindles down in this century. So interest in nuclear energy is still alive, but whether it will be a major source of future energy is still open to debate.

 INTERVIEW

With the permission of the National Energy Education Development (NEED) Project and Dr. Charles Ferguson, we have included a comprehensive career article in the nuclear energy field.

Green Advocate: Charles Ferguson, Philip D. Reed Senior Fellow for Science and Technology, Council on Foreign Relations, Washington, D.C.

Charles Ferguson holds a BS with distinction in physics from the U.S. Naval Academy, a master's degree equivalent in nuclear engineering from the Naval Nuclear Power School, and a PhD in physics from Boston University. In this interview Charles shares his unique and diverse involvement in the nuclear industry.

How or why did you choose to work in the nuclear industry?

I entered the nuclear industry through my training as a nuclear engineering officer in the Navy. I chose the nuclear submarine service because I thought it would be the most intellectually challenging service within the Navy. I served as a nuclear engineering officer of the watch on a ballistic missile submarine during the last years of the Cold War. As a watch officer, I was responsible for a 12-person crew that operated the power and propulsion systems of the submarine. As a result of the political changes happening at the end of the Cold War, I became very interested in nuclear arms control. These experiences about 20 years ago transformed my life and led to my current career. While the nuclear navy has been a gateway for many people to work in the commercial nuclear power industry, I am not currently working in this industry. Instead, I am working as an analyst who assesses policy options for nuclear power and examines where the industry appears to be headed. In addition, I research more effective ways for stopping the proliferation of nuclear weapons to countries and to terrorist groups. Thus, I work on public policy development on the peaceful and military sides of nuclear energy.

What subjects, courses, internships, or special training were instrumental in helping you gain your current position?

At the U.S. Naval Academy I majored in physics. When I graduated from the academy, I worked for a few months at the Los Alamos National Laboratory, which is one of the nuclear weapons laboratories. After this experience, I studied nuclear engineering at the Naval Nuclear Power School. I then served on a nuclear-powered submarine where I qualified as an engineering watch officer. After leaving

Dr. Charles Ferguson is a Philip D. Reed Senior Fellow for Science and Technology, Council on Foreign Relations, Washington, DC. (Courtesy Charles Ferguson)

the Navy, I decided to become a physics professor and earned my doctorate in physics at Boston University. However, my interest in nuclear nonproliferation and energy issues convinced me to leave the academic physics job track and take a job working on nuclear policy. The analytic training of my physics degree program has been helpful in my current policy work. One of the most influential jobs I had was working as a physical scientist on nuclear safety issues at the U.S. Department of State.

Would you follow the same career path again?

I have had a nontraditional career path that I could not have predicted more than 20 years ago when I started down it. I believe that if I had to start all over again, I probably would follow a similar path. My current career as a nuclear policy analyst has been very satisfying.

What do you think of the nuclear industry now that you work in it?

Because I analyze the industry rather than work in it, I have had the type of job in which I can step back and examine the big picture of the industry. On balance, the industry has made significant contributions to strengthening energy security and to countering climate change. However, there are still significant risks including the potential for terrorist attacks on nuclear facilities, the possibility that some countries may exploit the peaceful nuclear fuel cycle for military purposes, and the legacy of nuclear waste.

What opportunities or exciting experiences have you had in your career?

As part of my work, I have traveled to more than two dozen countries, including China, North Korea, Russia, Saudi Arabia, and Turkmenistan. When I worked at the State Department, I had the exciting experience of helping to secure radioactive materials that could fuel "dirty bombs" and to negotiate a treaty with Russia and other European countries to clean up nuclear waste in northwest Russia.

> **What is the most rewarding part of your job?**
>
> The most rewarding part is when I have had some influence on public policy through my writings, testimonies to Congress, or briefings to government officials. For example, a recommendation I made at a congressional hearing in 2007 led to the creation of a government program to improve the security of radioactive materials used in the nuclear industry and other industries.
>
> **What advice can you give to a young person considering a career in the nuclear industry?**
>
> Be intellectually curious about many things. Read deeply about the nuclear field but also read broadly about a range of subjects, especially science, politics, the environment, and history. Be disciplined in your studies and always strive to do your best.

BOOKS AND OTHER READING MATERIALS

Energy in Brief. *Nuclear: What Is the State of the U.S. Nuclear Industry?* Washington, D.C. http://www.tonto.eia.doe.gov.

Energy Information Administration. *Uranium (Nuclear) Basics.* Rockville, MD: U.S. Nuclear Regulatory Commission, 2008.

Energy Information Administration. *Uranium (Nuclear): Nuclear Power Plants.* Washington, D.C.: U.S. Department of Energy, 2008.

Hodgson, Peter E. *Nuclear Power, Energy and the Environment.* London: Imperial College Press, 1999.

Murray, Raymond L. *Nuclear Energy: An Introduction to the Concepts, Systems, and Applications of Nuclear Processes.* Burlington, MA: Butterworth-Heinemann, 1993.

Ramsey, Charles B., and Mohammad Modarres. *Commercial Nuclear Power: Assuring Safety for the Future.* Hoboken, NJ: Wiley, 1998.

Winnacker, Karl. *Nuclear Energy in Germany.* La Grange Park, IL: American Nuclear Society, 1979.

Wolfson, Richard. *Nuclear Choices: A Citizen's Guide to Nuclear Technology.* Rev. ed. Cambridge, MA: MIT Press, 1993.

SOMETHING TO DO

1. Both the Three Mile Island nuclear plant in Pennsylvania and the Chernobyl nuclear facility in Ukraine suffered meltdowns in their nuclear reactors. Although the accident at the Pennsylvania plant

was eventually controlled, the Chernobyl accident became a major disaster to the community and the environment. Compare the construction of the two plants, the reasons for the meltdowns, the management of the accidents, and the procedures in place for disaster control. You can use the following web sites for some of your research: www.loe.org; www.world-nuclear.org; and www.nrc.gov.
2. Contact your local energy companies and find out which types of reactors are used in your areas (light water reactor, boiling water reactor, and fast breeder reactor) and what their power ratings are. How do they dispose of their waste?

WEB SITES

The following Web sites, although not inclusive, include government and nongovernmental organizations.

www.nrc.gov

The U.S. Nuclear Regulatory Commission is headed by five commissioners appointed by the president and confirmed by the Senate for five-year terms. The commission formulates policies, develops regulations governing nuclear reactor and nuclear material safety, issues orders to licensees, and adjudicates legal matters.

www.ne.doe.gov

The Office of Nuclear Energy promotes nuclear power as a resource capable of meeting the nation's energy, environmental, and national security needs by resolving technical and regulatory barriers through research, development, and demonstration.

www.nirs.org

The Nuclear Information and Resource Service (NIRS) is an information and networking center for citizens and environmental activists concerned about nuclear power, radioactive waste, radiation, and sustainable energy issues.

www.ans.org

The core purpose of the American Nuclear Society is to promote the awareness and understanding of the application of nuclear science and technology.

www.nei.org

Nuclear Energy Institute's objective is to ensure the formation of policies that promote the beneficial uses of nuclear energy and technologies in the United States and around the world.

VIDEOS

The following video and audio selections are suggested to enhance your understanding of nuclear energy topics and issues. The author has made a consistent effort to include up-to-date Web sites. However, over time, some Web sites may move or no longer be available.

Viewing some of these videos may require special software called plug-ins. Therefore, you may need to download that software to view the videos. You also may need to upgrade your player to the most current version.

Nuclear Power—How It Works: A video presentation showing how the atom is used to generate electricity. To learn how 8×1-inch fuel pellets can generate enough energy to power a house for a whole year, go to http://www.youtube.com/watch?v=fjgdgAhOzXQ&feature=related (03:08 minutes).

How a Nuclear Pressurized Water Nuclear Reactor Works: A comprehensive video describing the three circuits involved in producing energy from U-235 pellets. To get to the heart of the matter, go to http://www.youtube.com/watch?v=u0VjHg0juz4&feature=related (06:03 minutes).

Nuclear Power—France: France meets nearly 80 percent of its energy needs with its 59 nuclear reactors, in a country the size of Texas. To learn more, go to http://www.youtube.com/watch?v=i-rKBrs7kYE&feature=related (10:00 minutes).

Nuclear Power—Australia: When it produces a large portion of the world's uranium, why is Australia resisting nuclear energy and 80 percent dependent on the carbon economy? For more, view this video: http://www.youtube.com/watch?v=mLPu_xPD8Qo&feature=fvw (10:09 minutes).

BOOKS AND OTHER READING MATERIALS

VOLUME 1: OIL, NATURAL GAS, COAL, AND NUCLEAR

American Petroleum Institute. *Natural Gas Supply and Demand.* http://www.api.org.

Boyle, Godfrey, ed. *Renewable Energy.* Oxford, UK: Oxford University Press, 2004.

Burns, Shirley Stewart. *Bringing Down the Mountains: The Impact of Mountaintop Removal on Southern West Virginia Communities.* Morgantown: West Virginia University Press, 2007.

Energy in Brief. *Nuclear: What Is the State of the U.S. Nuclear Industry?* http://www.tonto.eia.doe.gov.

Energy Information Administration. *Uranium (Nuclear) Basics.* Rockville, MD: U.S. Nuclear Regulatory Commission, 2008.

Energy Information Administration. *Uranium (Nuclear): Nuclear Power Plants.*

Freese, Barbara. *Coal: A Human History.* Cambridge, MA: Perseus Book Group, 2003.

Graham, Ian. *Fossil Fuels: A Resource Our World Depends Upon.* Chicago: Heinemann Library, 2005.

Nakaya, Andrea, ed. *Oil: Opposing Viewpoints.* San Diego, CA: Greenhouse Press, 2006.

Natural Gas Supply Association. *Natural Gas and the Environment.* www.naturalgas.org.

Richard, Julie. *Fossil Fuels.* North Mankato, MN: Smart Apple Media, 2003.

Riddle, John. *Coal Power of the Future.* New York: Rosen Publishing, 2003.

Sietz, John L. *Global Issues: An Introduction.* Malden, MA: Blackwell, 2002.

Smil, Vaclav. *Oil: Beginner's Guide.* Oxford, UK: One World Publications, 2008.

VOLUME 2: SOLAR ENERGY AND HYDROGEN FUEL CELLS

Craddock, David. *Renewable Energy Made Easy: Free Energy from Solar, Wind, Hydropower, and other Alternative Energy Sources.* Ocala, FL: Atlantic Publishing, 2008.

Ewing, Rex A. *Got Sun? Go Solar: Harness Nature's Free Energy to Heat and Power Your Grid-Tied Home.* Masonville, CO: PixyJack Press, 2009.

Harper, Gavin D. J. *Solar Energy Projects for the Evil Genius.* New York: McGraw-Hill, 2007.

Haugen, David M., ed. *Hydrogen.* Detroit, MI: Greenhaven Press, 2006.

Hayhurst, Chris. *Hydrogen Power: New Ways of Turning Fuel Cells into Energy.* New York: Rosen, 2003.

Jones, Susan. *Solar Power of the Future: New Ways of Turning Sunlight into Energy.* New York: Rosen, 2002.

Kachadorian, James. *The Passive Solar House.* White River Junction, VT: Chelsea Green, 2006.

Kryza, Frank. *The Power of Light: The Epic Story of Man's Quest to Harness the Sun.* New York: McGraw-Hill, 2003.

Oxlade, Chris. *Solar Energy.* Chicago: Heinemann Library, 2008.

Pieper, Adi. *The Easy Guide to Solar Electric.* Santa Fe, NM: ADI Solar, 2001.

Ramsey, Dan, with David Hughes. *The Complete Idiot's Guide to Solar Power for Your Home.* New York: Alpha Books, 2007.

Smith, Trevor. *Renewable Energy Resources.* Mankato, MN: Weigh Publishers, 2003.

Solway, Andrew. *Hydrogen Fuel.* Pleasantville, NY: Gareth Stevens, 2008.

Vaitheeswaran, Vijay V. *Power to the People: How the Coming Energy Revolution Will Transform an Industry, Change Our Lives, and Maybe Even Save the Planet.* New York: Farrar, Straus and Giroux, 2003.

Walker, Niki. *Hydrogen: Running on Water.* St. Catharines, ON: Crabtree, 2007.

VOLUME 3: WIND ENERGY, OCEANIC ENERGY, AND HYDROPOWER

American Wind Energy Association. *Wind Web Tutorial.* http://www.awea.org.

Energy Resources: Tidal Power. http://www.clara.net.

Gasch, Robert. *Wind Power Plants: Fundamentals, Design, Construction and Operation.* London: Earthscan, 2004.

Gipe, Paul. *Wind Power: Renewable Energy for Home, Farm or Business.* White River Junction, VT: Chelsea Green, 2004.

Koller, Julia. *Offshore Wind Energy.* New York: Springer, 2006.

Matthew, Sathyajith. *Wind Energy Fundamentals.* New York: Springer, 2006.

Morris, Neil. *Water Power.* North Mankato, MN: Apple Media, 2006.

National Renewable Energy Laboratory and U.S. Department of Energy. *Wind Energy Information Guide.* Honolulu, HI: University Press of the Pacific, 2005.

Pasqualetti, Martin. *Wind Power in View: Energy Landscapes in a Crowded World.* San Diego, CA: Academic Press, 2002.

Renewable Energy, UK. *Introduction to Tidal Power.* http://www.reuk.co.uk.

Renewable Energy, UK. *Severn Barrage Tidal Power.* http://www.reuk.co.uk.

Szarka, Joseph. *Wind Power in Europe.* New York: Palgrave MacMillan, 2007.

U.S. Department of Energy. *How a Microhydropower System Works.* http://www.energy.gov/forresearchers.

U.S. Department of Interior and the U.S. Geological Survey. *Hydroelectric Power: How It Works.* http://www.library.usgs.gov.

Whitcomb, Robert. *Cape Wind . . . and the Battle for Our Energy Future on Nantucket Sound.* New York: Palgrave MacMillan, 2007.

VOLUME 4: GEOTHERMAL AND BIOMASS ENERGY

Armentrout, David, and Patricia Armentrout. *Biofuels*. Vero Beach, FL: Rourke, 2009.

Garza, Amanda de la, ed. *Biomass: Energy from Plants and Animals*. Detroit, MI: Greenhaven Press, 2007.

Haugen, David M., ed. *Fueling the Future / Biomass*. Detroit, MI: Greenhaven Press, 2007.

Hayhurst, Chris. *Biofuel Power of the Future: New Ways of Turning Organic Matter into Energy*. New York: Rosen, 2002.

Kemp, William H. *The Renewable Energy Handbook: A Guide to Rural Independence, Off-Grid and Sustainable Living*. Tamworth, Ontario: Aztext Press, 2005.

Morris, Neil. *Biomass Power*. North Mankato, MN: Smart Apple Media, 2007.

Morris, Neil. *Geothermal Power*. North Mankato, MN: Smart Apple Media, 2007.

Orr, Tamra. *Geothermal Energy*. Ann Arbor, MI: Cherry Lake Publishing, 2008.

Pahl, Greg. *Biodiesel: Growing a New Energy Economy*. White River Junction, VT: Chelsea Green, 2005.

Povey, Karen D. *Biofuels*. San Diego, CA: KidHaven Press, 2007.

Saunders, N. *Geothermal Energy*. Pleasantville, NY: Gareth Stevens, 2008.

Savage, Lorraine, ed. *Geothermal Power*. Detroit, MI: Greenhaven Press, 2007.

Sherman, Josepha. *Geothermal Power*. Mankato, MN: Capstone Press, 2004.

Tabak, John. *Biofuels*. New York: Facts on File, 2009.

Walker, Niki. *Biomass: Fueling Change*. New York: Crabtree, 2007.

VOLUME 5: ENERGY EFFICIENCY, CONSERVATION, AND SUSTAINABILITY

Bauer, Seth, ed. *Green Guide*. Washington, DC: National Geographic, 2008.

Chiras, Dan. *The Homeowner's Guide to Renewable Energy*. Gabriola Island, BC: New Society, 2006.

Edwards, Andre. *The Sustainability Revolution*. Gabriola Island, BC: New Society, 2005.

Freeman, S. David. *Winning Our Energy Independence.* Salt Lake City, UT: Gibbs Smith, 2007.

Gore, Al. *An Inconvenient Truth.* Emmaus, PA: Rodale Press, 2006.

Grant, Tim, and Gail Littlejohn. *Greening School Grounds.* Gabriola Island, BC: New Society, 2001.

Krigger, John, and Chris Dorsi. *The Homeowner's Handbook to Energy Efficiency.* Helena, MT: Saturn Resource Management, 2008.

Osmundson, Theodore. *Roof Gardens: History, Design and Construction.* New York: Norton, 2000.

Riley, Trish. *Guide to Green Living.* New York: Alpha-Penguin, 2007.

Roberts, Jennifer. *Good Green Homes.* Layton, UT: Gibbs Smith, 2003.

Schaeffer, John, ed. *Real Goods Solar Living Source Book.* Hopland, CA: Real Goods Trading, 2007.

Schor, Juliet B., and Betsy Taylor. *Sustainable Planet: Solutions for the Twenty-First Century.* Boston: Beacon Press, 2002.

Trask, Crissy. *It's Easy Being Green.* Salt Lake City, UT: Gibbs Smith, 2006.

U.S. Department of Energy. *A Place in the Sun: Solar Buildings.* Merryfield, VA: EERE Clearing House, 2005.

U.S. Green Building Council. *Meet the USGBC: Mission Statement.* http://www.usgbc.org.

GOVERNMENT AND NONGOVERNMENTAL ORGANIZATION WEB SITES

Agency for Toxic Substances and Diseases: www.atsdr.cdc.gov/contacts.html
American Gas Association: www.aga.org
American Nuclear Society: www.ans.org
American Oceans Campaign: www.americanoceans.org
American Petroleum Institute: www.api.org
American Solar Energy Society: www.ases.org
American Wind Energy Association: www.awea.org
Center for Renewable Energy and Sustainable Technology (CREST), Solar Energy Research and Education Foundation: solstice.crest.org/
Clean Air Council (CAC): www.libertynet.org/~cleanair/
Coal Age Magazine: coalage.com
Coalition for Economically Responsible Economies (CERES): www.ceres.org
Electric Vehicle Association of the Americas: www.evaa.org
Environmental Defense Fund: www.edf.org
Federal Emergency and Management Agency (FEMA): www.fema.gov
Hazard Ranking System: www.epa.gov/superfund/programs/npl_hrs/hrsint.htm
Hydrogen InfoNet: /www.eren.doe.gov/hydrogen/infonet.html

International Atomic Energy Commission: www.iaea.org
International Centre for Antarctic Information and Research: www.icair.iac.org.nz
International Council for Local Environmental Initiatives (ICLEI): www.iclei.org
Los Alamos National Laboratory: www.lanl.gov/wvu.edu/news/nsamd.html
National Ocean and Atmospheric Administration and Divisions: www.noaa.gov/
National Renewable Energy Laboratory: www.nrel.gov/
National Research Center for Coal and Energy, West Virginia University: www.nrcce.wvu.edu
Natural Resources Conservation Service: www.nrcs.usda.gov
National Science Foundation (NSF): www.nsf.gov/crssprgm/nano/
National Weather Service: www.nws.noaa.gov
Noise Pollution Clearinghouse: www.nonoise.org
North Sea Commission: www.northsea.org
Nuclear Energy Institute: www.nei.org
Nuclear Regulatory Commission: www.nrc.gov
Office of Surface Mining: www.osmre.gov
Organization of Petroleum Exporting Countries (OPEC): www.opec.org
Ozone Action: www.ozone.org
Resources for the Future (RFF): www.sandia.gov/
Superfund: www.epa.gov/superfund
Union of Concerned Scientists: www.ucsusa.org
United Nations Environment Programme: www.unep.org
United Nations Food and Agricultural Organization (FAO): www.fao.org
United Nations Man and the Biosphere Programme (UNMAB): www.mabnet.org
United States Bureau of Reclamation, Hydropower Information: www.usbr.gov/power/edu/edu.html
United States Department of Agriculture (USDA): www.usda.gov
United States Department of Defense (DOD): www.defenselink.mil/
United States Department of Education: www.ed.gov/index.jhtml
United States Department of Energy: www.energy.gov/index.htm
United States Department of the Interior: www.doi.gov

United States Environmental Protection Agency (EPA): www.epa.gov
United States Geological Survey (USGS): www.usgs.gov
United States Geological Survey (USGS), Geology Research: geology.usgs.gov/index.shtml
World Conservation Monitoring Centre (WCMC): www.wcmc.org.uk
World Resources Institute: www.wri.org/wri/biodiv; e-mail: info@wri.org

ENERGY DATA

The eight tables in this section include information about the United States and the world's consumption of nonrenewable and renewable energy sources, and how various sectors use energy. These kinds of statistics are vital to economists, energy theorists, policymakers, engineers, and environmentalists for predicting future energy demands and assessing to what extent the world's remaining resources can meet those energy needs. In addition, such data show which countries consume the most energy, produce the most energy, and contribute the most pollution due to energy intake—all valuable factors to take into consideration as a global economy, waning natural resources, and growing world population require increasing worldwide cooperation when it comes to energy policy. Due to the pervasiveness of energy in our everyday lives, these types of data are important even to citizens who do not directly work for the energy sector.

Table 1: Primary Energy Consumption by Source, 1949–2008
Data on U.S. energy use, listing the annual consumption amounts by individual energy sources and categorized into renewable and nonrenewable categories.

Table 2: Renewable Energy Production and Consumption by Primary Energy Source, 1949–2008
Data on U.S. renewable energy production and consumption, divided by source.

Table 3: Energy Consumption by Sector, 1949–2008
Energy use statistics of four main sectors in the United States: residential, commercial, industrial, and transportation.

Table 4: Household End Uses: Fuel Types and Appliances, Selected Years, 1978–2005
Energy consumption in the U.S. housing sector, including appliance-specific energy use and energy sources used for household heating and cooling purposes.

Table 5: World Primary Energy Consumption by Region, 1997–2006
Total energy use by world region and country.

Table 6: World Crude Oil and Natural Gas Reserves, January 1, 2008
Amount of oil and natural gas reserves available as of 2008 by world region and country.

Table 7: World Recoverable Reserves of Coal, 2005
Amount of coal reserves technologically and economically feasible to recover as of 2005, listed by region, country, and type of coal.

Table 8: World Carbon Dioxide Emissions from Energy Consumption, 1997–2006
Data listing the amount of carbon dioxide emitted by each world region and country.

TABLE 1 Primary Energy Consumption by Source, 1949–2008 (Billion Btu)

	Fossil Fuels					Nuclear Electric Power	Renewable Energy[a]
Year	Coal	Coal Coke Net Imports[b]	Natural Gas[c]	Petroleum[d]	Total		Hydro-electric Power[e]
1949	11,980,905	−6,671	5,145,142	11,882,722	29,002,099	0	1,424,722
1950	12,347,109	992	5,968,371	13,315,484	31,631,956	0	1,415,411
1951	12,552,996	−21,452	7,048,518	14,428,043	34,008,105	0	1,423,795
1952	11,306,479	−11,879	7,549,621	14,955,682	33,799,903	0	1,465,812
1953	11,372,684	−9,002	7,906,645	15,555,829	34,826,156	0	1,412,859
1954	9,714,667	−6,746	8,330,202	15,839,176	33,877,300	0	1,359,772
1955	11,167,259	−10,044	8,997,935	17,254,955	37,410,105	0	1,359,844
1956	11,349,723	−13,020	9,613,975	17,937,473	38,888,151	0	1,434,711
1957	10,820,631	−17,459	10,190,753	17,931,667	38,925,592	112	1,515,613
1958	9,533,287	−6,721	10,663,199	18,526,937	38,716,702	1,915	1,591,967
1959	9,518,353	−8,358	11,717,422	19,322,650	40,550,068	2,187	1,548,465
1960	9,837,785	−5,630	12,385,366	19,919,230	42,136,751	6,026	1,607,975
1961	9,623,351	−7,886	12,926,392	20,216,387	42,758,243	19,678	1,656,463
1962	9,906,454	−5,506	13,730,841	21,048,981	44,680,770	26,394	1,816,141
1963	10,412,538	−7,390	14,403,306	21,700,828	46,509,283	38,147	1,771,355
1964	10,964,385	−10,441	15,287,850	22,301,257	48,543,050	39,819	1,886,314
1965	11,580,608	−18,451	15,768,667	23,245,680	50,576,504	43,164	2,059,077
1966	12,143,080	−24,949	16,995,332	24,400,523	53,513,987	64,158	2,061,519
1967	11,913,750	−15,326	17,944,788	25,283,661	55,126,873	88,456	2,346,664
1968	12,330,677	−17,310	19,209,656	26,979,447	58,502,470	141,534	2,348,629
1969	12,381,540	−36,109	20,677,984	28,338,336	61,361,751	153,722	2,647,983
1970	12,264,528	−57,660	21,794,707	29,520,695	63,522,269	239,347	2,633,547
1971	11,598,411	−33,108	22,469,052	30,561,290	64,595,645	412,939	2,824,151
1972	12,076,917	−25,966	22,698,190	32,946,738	67,695,880	583,752	2,863,865
1973	12,971,490	−7,465	22,512,399	34,839,926	70,316,351	910,177	2,861,448
1974	12,662,878	56,098	21,732,488	33,454,627	67,906,091	1,272,083	3,176,580
1975	12,662,786	13,541	19,947,883	32,730,587	65,354,796	1,899,798	3,154,607
1976	13,584,067	−99	20,345,426	35,174,688	69,104,082	2,111,121	2,976,265
1977	13,922,103	14,582	19,930,513	37,122,168	70,989,367	2,701,762	2,333,252
1978	13,765,575	124,719	20,000,400	37,965,295	71,855,989	3,024,126	2,936,983
1979	15,039,586	62,843	20,665,817	37,123,381	72,891,627	2,775,827	2,930,686
1980	15,422,809	−35,018	20,235,459	34,202,356	69,825,607	2,739,169	2,900,144
1981	15,907,526	−15,946	19,747,309	31,931,050 [R]	67,569,939 [R]	3,007,589	2,757,968
1982	15,321,581	−21,650	18,356,222	30,231,608 [R]	63,887,761 [R]	3,131,148	3,265,558
1983	15,894,442	−15,624	17,220,836	30,053,921 [R]	63,153,575 [R]	3,202,549	3,527,260
1984	17,070,622	−11,482	18,393,613	31,051,327	66,504,079	3,552,531	3,385,811
1985	17,478,428	−13,491	17,703,482	30,922,149 [R]	66,090,567 [R]	4,075,563	2,970,192
1986	17,260,405	−16,740	16,591,364	32,196,080	66,031,109	4,380,109	3,071,179
1987	18,008,451	8,630	17,639,801	32,865,053 [R]	68,521,935 [R]	4,753,933	2,634,508
1988	18,846,312	39,556	18,448,393	34,221,992	71,556,253	5,586,968	2,334,265
1989	19,069,762	30,405	19,601,689	34,211,114	72,912,970	5,602,161	2,837,263
1990	19,172,635	4,786	19,603,168	33,552,534	72,333,123	6,104,350	3,046,391
1991	18,991,670	9,697	20,032,957	32,845,361	71,879,686 [R]	6,422,132	3,015,943
1992	19,122,471	34,621	20,713,632	33,526,585 [R]	73,397,310 [R]	6,479,206	2,617,436
1993	19,835,148	27,106	21,228,902	33,744,490 [R]	74,835,647 [R]	6,410,499	2,891,613
1994	19,909,463	58,330	21,728,065	34,561,665	76,257,523	6,693,877	2,683,457
1995	20,088,727	61,058	22,671,138	34,436,967 [R]	77,257,890 [R]	7,075,436	3,205,307
1996	21,001,914	22,816	23,084,647	35,673,290 [R]	79,782,668 [R]	7,086,674	3,589,656

Renewable Energy[a]					Electricity Net Imports[b]	Total
Geothermal	Solar/PV	Wind	Biomass	Total		
NA	NA	NA	1,549,262	2,973,984	5,420	31,981,503
NA	NA	NA	1,562,307	2,977,718	6,094	34,615,768
NA	NA	NA	1,534,669	2,958,464	7,461	36,974,030
NA	NA	NA	1,474,369	2,940,181	7,740	36,747,825
NA	NA	NA	1,418,601	2,831,460	6,852	37,664,468
NA	NA	NA	1,394,327	2,754,099	7,983	36,639,382
NA	NA	NA	1,424,143	2,783,907	13,879	40,207,971
NA	NA	NA	1,415,871	2,850,582	15,519	41,754,252
NA	NA	NA	1,333,581	2,849,194	12,288	41,787,186
NA	NA	NA	1,323,123	2,915,090	11,320	41,645,028
NA	NA	NA	1,352,874	2,901,339	12,127	43,465,722
774	NA	NA	1,319,870	2,928,619	15,474	45,086,870
2,181	NA	NA	1,294,762	2,953,406	7,689	45,739,017
2,331	NA	NA	1,300,242	3,118,714	1,829	47,827,707
3,726	NA	NA	1,323,316	3,098,396	334	49,646,160
4,520	NA	NA	1,336,802	3,227,637	6,671	51,817,177
4,197	NA	NA	1,334,761	3,398,036	−482	54,017,221
4,170	NA	NA	1,368,985	3,434,674	3,725	57,016,544
6,886	NA	NA	1,340,249	3,693,799	−1,020	58,908,107
9,416	NA	NA	1,419,495	3,777,541	−2,152	62,419,392
13,281	NA	NA	1,440,487	4,101,751	3,656	65,620,879
11,347	NA	NA	1,430,962	4,075,857	6,688	67,844,161
11,862	NA	NA	1,432,323	4,268,335	12,046	69,288,965
31,479	NA	NA	1,503,065	4,398,409	26,227	72,704,267
42,605	NA	NA	1,529,068	4,433,121	48,715	75,708,364
53,158	NA	NA	1,539,657	4,769,395	43,311	73,990,880
70,153	NA	NA	1,498,734	4,723,494	21,103	71,999,191
78,154	NA	NA	1,713,373	4,767,792	29,378	76,012,373
77,418	NA	NA	1,838,332	4,249,002	59,422	77,999,554
64,350	NA	NA	2,037,605	5,038,938	67,318	79,986,371
83,788	NA	NA	2,151,906	5,166,379	69,381	80,903,214
109,776	NA	NA	2,475,500	5,485,420	71,399	78,121,594
123,043	NA	NA	2,596,542 [R]	5,477,554 [R]	113,406	76,168,488 [R]
104,746	NA	NA	2,664,154 [R]	6,034,459 [R]	100,026	73,153,394 [R]
129,339	NA	28	2,905,703 [R]	6,562,330 [R]	120,547	73,039,001 [R]
164,896	55	68	2,972,697 [R]	6,523,526 [R]	135,323	76,715,459 [R]
198,282	111	60	3,018,134 [R]	6,186,780 [R]	139,655	76,492,565 [R]
219,178	147	44	2,934,280 [R]	6,224,827 [R]	122,481	76,758,526 [R]
229,119	109	37	2,877,388 [R]	5,741,161 [R]	158,101	79,175,130 [R]
217,290	94	9	3,018,580 [R]	5,570,238 [R]	108,399	82,821,858 [R]
317,163	55,291	22,033	3,161,916 [R]	6,393,667 [R]	37,450	84,946,248 [R]
335,801	59,718	29,007	2,737,372 [R]	6,208,290 [R]	7,888	84,653,651 [R]
346,247	62,688	30,796	2,784,410 [R]	6,240,085 [R]	66,965	84,608,869 [R]
349,309	63,886	29,863	2,934,637 [R]	5,995,131 [R]	86,733	85,958,380 [R]
363,716	66,458	30,987	2,911,622 [R]	6,264,397 [R]	94,910	87,605,453 [R]
338,108	68,548	35,560	3,031,380 [R]	6,157,054 [R]	152,937	89,261,391 [R]
293,893	69,857	32,630	3,105,220 [R]	6,706,907 [R]	133,856	91,174,089 [R]
315,529	70,833	33,440	3,159,720 [R]	7,169,179 [R]	137,144	94,175,664 [R]

TABLE 1 (*Continued*)

Year	Fossil Fuels					Nuclear Electric Power	Renewable Energy[a]
	Coal	Coal Coke Net Imports[b]	Natural Gas[c]	Petroleum[d]	Total		Hydro-electric Power[e]
1997	21,445,411	46,450	23,222,718	36,159,835 [R]	80,874,414 [R]	6,596,992	3,640,458
1998	21,655,744	67,084	22,830,226	36,816,619	81,369,672	7,067,809	3,297,054
1999	21,622,544	57,685	22,909,227	37,838,081 [R]	82,427,536 [R]	7,610,256	3,267,575
2000	22,579,528	65,348	23,823,978	38,264,303 [R]	84,733,157 [R]	7,862,349	2,811,116
2001	21,914,268	29,264	22,772,558	38,186,476 [R]	82,902,566 [R]	8,032,697	2,241,858
2002	21,903,989	60,760	23,558,419	38,226,666 [R]	83,749,834 [R]	8,143,089	2,689,017
2003	22,320,928	50,518	22,897,268	38,809,183 [R]	84,077,896 [R]	7,958,858	2,824,533
2004	22,466,195	137,739	22,931,481	40,294,351	85,829,766	8,221,985	2,690,078
2005	22,796,543	44,194	22,583,385	40,393,325	85,817,446	8,160,028	2,702,942
2006	22,447,160	60,810	22,223,903 [R]	39,958,151 [R]	84,690,024 [R]	8,213,839	2,869,035
2007	22,749,466 [R]	25,197	23,627,629 [R]	39,773,213 [R]	86,175,506 [R]	8,457,783 [R]	2,446,389 [R]
2008 P	22,420,827	40,771	23,837,695	37,136,675	83,435,968	8,455,236	2,452,073

[a] Most data are estimates.

[b] Net imports equal imports minus exports. Minus sign indicates exports are greater than imports.

[c] Natural gas only; excludes supplemental gaseous fuels.

[d] Petroleum products supplied, including natural gas plant liquids and crude oil burned as fuel. Does not include the fuel ethanol portion of motor gasoline—fuel ethanol is included in "Biomass."

[e] Conventional hydroelectric power.

R = Revised. P = Preliminary. NA = Not available. (s) = Less than 0.0005 and greater than −0.0005 quadrillion Btu.

Source: U.S. Energy Information Administration Annual Energy Review 2009.

Renewable Energy[a]						Electricity Net Imports[b]	Total	
Geothermal	Solar/PV	Wind	Biomass		Total			
324,959	70,237	33,581	3,108,968 [R]	7,178,202 [R]		116,203	94,765,811	[R]
328,303	69,787	30,853	2,931,592	6,657,589 [R]		88,224	95,183,293	[R]
330,919	68,793	45,894	2,967,555 [R]	6,680,737 [R]		98,924	96,817,452	[R]
316,796	66,388	57,057	3,013,038 [R]	6,264,394 [R]		115,199	98,975,100	[R]
311,264	65,454	69,617	2,627,476 [R]	5,315,670 [R]		75,156	96,326,089	[R]
328,308	64,391	105,334	2,706,745 [R]	5,893,795 [R]		71,595	97,858,314	[R]
330,554	63,620	114,571	2,816,604 [R]	6,149,881 [R]		21,905 [R]	98,208,541	[R]
341,082	64,500	141,749	3,022,866 [R]	6,260,276 [R]		38,597	100,350,624	[R]
342,576	66,130	178,088	3,133,146 [R]	6,422,883 [R]		84,401 [R]	100,484,758	[R]
342,876	72,222	263,738	3,360,613 [R]	6,908,484 [R]		62,849	99,875,196	
348,730 [R]	80,943 [R]	340,503 [R]	3,597,370 [R]	6,813,935 [R]		106,632	101,553,855	[R]
358,497	91,003	514,224	3,884,252	7,300,048		112,381	99,303,634	

TABLE 2 Renewable Energy Production and Consumption by Primary Energy Source, 1949–2008 (Billion Btu)

	Production[a]			Consumption		
	Biomass		Total Renewable	Hydro-electric		
Year	Biofuels[b]	Total[c]	Energy[d]	Power[e]	Geo-thermal[f]	Solar/PV[g]
1949	NA	1,549,262	2,973,984	1,424,722	NA	NA
1950	NA	1,562,307	2,977,718	1,415,411	NA	NA
1951	NA	1,534,669	2,958,464	1,423,795	NA	NA
1952	NA	1,474,369	2,940,181	1,465,812	NA	NA
1953	NA	1,418,601	2,831,460	1,412,859	NA	NA
1954	NA	1,394,327	2,754,099	1,359,772	NA	NA
1955	NA	1,424,143	2,783,987	1,359,844	NA	NA
1956	NA	1,415,871	2,850,582	1,434,711	NA	NA
1957	NA	1,333,581	2,849,194	1,515,613	NA	NA
1958	NA	1,323,123	2,915,090	1,591,967	NA	NA
1959	NA	1,352,874	2,901,339	1,548,465	NA	NA
1960	NA	1,319,870	2,928,619	1,607,975	774	NA
1961	NA	1,294,762	2,953,406	1,656,463	2,181	NA
1962	NA	1,300,242	3,118,714	1,816,141	2,331	NA
1963	NA	1,323,316	3,098,396	1,771,355	3,726	NA
1964	NA	1,336,802	3,227,637	1,886,314	4,520	NA
1965	NA	1,334,761	3,398,036	2,059,077	4,197	NA
1966	NA	1,368,985	3,434,674	2,061,519	4,170	NA
1967	NA	1,340,249	3,693,799	2,346,664	6,886	NA
1968	NA	1,419,495	3,777,541	2,348,629	9,416	NA
1969	NA	1,440,487	4,101,751	2,647,983	13,281	NA
1970	NA	1,430,962	4,075,857	2,633,547	11,347	NA
1971	NA	1,432,323	4,268,335	2,824,151	11,862	NA
1972	NA	1,503,065	4,398,409	2,863,865	31,479	NA
1973	NA	1,529,068	4,433,121	2,861,448	42,605	NA
1974	NA	1,539,657	4,769,395	3,176,580	53,158	NA
1975	NA	1,498,734	4,723,494	3,154,607	70,153	NA
1976	NA	1,713,373	4,767,792	2,976,265	78,154	NA
1977	NA	1,838,332	4,249,002	2,333,252	77,418	NA
1978	NA	2,037,605	5,038,938	2,936,983	64,350	NA
1979	NA	2,151,906	5,166,379	2,930,686	83,788	NA
1980	NA	2,475,500	5,485,420	2,900,144	109,776	NA
1981	12,979 [R]	2,596,542 [R]	5,477,554 [R]	2,757,968	123,043	NA
1982	35,106 [R]	2,664,154 [R]	6,034,459 [R]	3,265,558	104,746	NA
1983	64,432 [R]	2,905,703 [R]	6,562,330 [R]	3,527,260	129,339	NA
1984	78,880 [R]	2,972,697 [R]	6,523,526 [R]	3,385,811	164,896	55
1985	95,052 [R]	3,018,134 [R]	6,186,780 [R]	2,970,192	198,282	111
1986	109,285 [R]	2,934,280 [R]	6,224,827 [R]	3,071,179	219,178	147
1987	125,229 [R]	2,877,388 [R]	5,741,161 [R]	2,634,508	229,119	109
1988	126,589 [R]	3,018,580 [R]	5,570,238 [R]	2,334,265	217,290	94
1989	127,936 [R]	3,161,916 [R]	6,393,667 [R]	2,837,263	317,163	55,291
1990	113,129 [R]	2,737,372 [R]	6,208,290 [R]	3,046,391	335,801	59,718
1991	130,612 [R]	2,784,410 [R]	6,240,085 [R]	3,015,943	346,247	62,688
1992	147,965 [R]	2,934,637 [R]	5,995,131 [R]	2,617,436	349,309	63,886
1993	172,792 [R]	2,911,902 [R]	6,264,676 [R]	2,891,613	363,716	66,458
1994	192,236 [R]	3,031,380 [R]	6,157,054 [R]	2,683,457	338,108	68,548
1995	201,773 [R]	3,103,118 [R]	6,704,805 [R]	3,205,307	293,893	69,857
1996	144,167 [R]	3,158,184 [R]	7,167,643 [R]	3,589,656	315,529	70,833

	Consumption					Total Renewable
		Biomass				
Wind[h]	Wood[i]	Waste[j]	Biofuels[k]		Total	Energy
NA	1,549,262	NA	NA		1,549,262	2,973,984
NA	1,562,307	NA	NA		1,562,307	2,977,718
NA	1,534,669	NA	NA		1,534,669	2,958,464
NA	1,474,369	NA	NA		1,474,369	2,940,181
NA	1,418,601	NA	NA		1,418,601	2,831,460
NA	1,394,327	NA	NA		1,394,327	2,754,099
NA	1,424,143	NA	NA		1,424,143	2,783,987
NA	1,415,871	NA	NA		1,415,871	2,850,582
NA	1,333,581	NA	NA		1,333,581	2,849,194
NA	1,323,123	NA	NA		1,323,123	2,915,090
NA	1,352,874	NA	NA		1,352,874	2,901,339
NA	1,319,870	NA	NA		1,319,870	2,928,619
NA	1,294,762	NA	NA		1,294,762	2,953,406
NA	1,300,242	NA	NA		1,300,242	3,118,714
NA	1,323,316	NA	NA		1,323,316	3,098,396
NA	1,336,802	NA	NA		1,336,802	3,227,637
NA	1,334,761	NA	NA		1,334,761	3,398,036
NA	1,368,985	NA	NA		1,368,985	3,434,674
NA	1,340,249	NA	NA		1,340,249	3,693,799
NA	1,419,495	NA	NA		1,419,495	3,777,541
NA	1,440,487	NA	NA		1,440,487	4,101,751
NA	1,428,649	2,313	NA		1,430,962	4,075,857
NA	1,430,229	2,094	NA		1,432,323	4,268,335
NA	1,500,992	2,073	NA		1,503,065	4,398,409
NA	1,527,012	2,056	NA		1,529,068	4,433,121
NA	1,537,755	1,902	NA		1,539,657	4,769,395
NA	1,496,928	1,806	NA		1,498,734	4,723,494
NA	1,711,484	1,889	NA		1,713,373	4,767,792
NA	1,836,524	1,808	NA		1,838,332	4,249,002
NA	2,036,150	1,455	NA		2,037,605	5,038,938
NA	2,149,854	2,052	NA		2,151,906	5,166,379
NA	2,473,861	1,639	NA		2,475,500	5,485,420
NA	2,495,563	88,000	12,979	[R]	2,596,542 [R]	5,477,554 [R]
NA	2,510,048	119,000	35,106	[R]	2,664,154 [R]	6,034,459 [R]
28	2,684,271	157,000	64,432	[R]	2,905,703 [R]	6,562,330 [R]
68	2,685,817	208,000	78,880	[R]	2,972,697 [R]	6,523,526 [R]
60	2,686,765	236,317	95,052	[R]	3,018,134 [R]	6,186,780 [R]
44	2,562,134	262,861	109,285	[R]	2,934,280 [R]	6,224,827 [R]
37	2,463,159	289,000	125,229	[R]	2,877,388 [R]	5,741,161 [R]
9	2,576,663	315,328	126,589	[R]	3,018,580 [R]	5,570,238 [R]
22,033	2,679,623	354,357	127,936	[R]	3,161,916 [R]	6,393,667 [R]
29,007	2,216,165	408,078	113,129	[R]	2,737,372 [R]	6,208,290 [R]
30,796	2,214,083	439,715	130,612	[R]	2,784,410 [R]	6,240,085 [R]
29,863	2,313,471	473,201	147,965	[R]	2,934,637 [R]	5,995,131 [R]
30,987	2,259,774	479,336 [R]	172,512	[R]	2,911,622 [R]	6,264,397 [R]
35,560	2,323,820	515,324	192,236		3,031,380 [R]	6,157,054 [R]
32,630	2,369,869	531,476 [R]	203,875		3,105,220 [R]	6,706,907 [R]
33,440	2,437,027	576,990	145,703	[R]	3,159,720 [R]	7,169,179 [R]

TABLE 2 (Continued)

Year	Production[a] Biomass Biofuels[b]		Production[a] Biomass Total[c]		Production[a] Total Renewable Energy[d]		Consumption Hydro-electric Power[e]		Consumption Geo-thermal[f]		Consumption Solar/PV[g]	
1997	190,117	[R]	3,111,710	[R]	7,180,944	[R]	3,640,458		324,959		70,237	
1998	206,606	[R]	2,933,061	[R]	6,659,058	[R]	3,297,054		328,303		69,787	
1999	215,111	[R]	2,969,434	[R]	6,682,616	[R]	3,267,575		330,919		68,793	
2000	237,904		3,010,419	[R]	6,261,775	[R]	2,811,116		316,796		66,388	
2001	259,624	[R]	2,629,331	[R]	5,317,524	[R]	2,241,858		311,264		65,454	
2002	314,379	[R]	2,711,668	[R]	5,898,718	[R]	2,689,017		328,308		64,391	
2003	411,484	[R]	2,814,871	[R]	6,148,149	[R]	2,824,533		330,554		63,620	
2004	500,262	[R]	3,010,557	[R]	6,247,966	[R]	2,690,078		341,082		64,500	
2005	580,572	[R]	3,120,142	[R]	6,409,879	[R]	2,702,942		342,576		66,130	
2006	743,069	[R]	3,309,026	[R]	6,856,897	[R]	2,869,035		342,876		72,222	
2007	1,010,932	[R]	3,583,444	[R]	6,800,009	[R]	2,446,389	[R]	348,730	[R]	80,943	[R]
2008P	1,428,745		3,899,915		7,315,711		2,452,073		358,497		91,003	

[a]Production equals consumption for all renewable energy sources except biofuels.

[b]Total biomass inputs to the production of fuel ethanol and biodiesel.

[c]Wood and wood-derived fuels, biomass waste, fuel ethanol, and biodiesel.

[d]Hydroelectric power, geothermal, solar/PV, wind, and biomass.

[e]Conventional hydroelectricity net generation (converted to Btu using the fossil-fueled plant's heat rate).

[f]Geothermal electricity net generation (converted to Btu using the geothermal energy plant's heat rate), and geothermal heat pump and direct-use energy.

[g]Solar thermal and photovoltaic electricity net generation (converted to Btu using the fossil-fueled plant's heat rate) and solar thermal direct-use energy.

[h]Wind electricity net generation (converted to Btu using the fossil-fueled plant's heat rate).

[i]Wood and wood-derived fuels.

[j]Municipal solid waste from biogenic sources, landfill gas, sludge waste, agricultural byproducts, and other biomass. Through 2000, also includes nonrenewable waste (municipal solid waste from non-biogenic sources and tire-derived fuels).

[k]Fuel ethanol and biodiesel consumption, plus losses and co-products from the production of fuel ethanol and biodiesel.

R = Revised. P = Preliminary. NA = Not available. (s) = Less than 0.5 trillion Btu.

Note: Totals may not equal sum of components as a result of independent rounding. For related information, see http://www.eia.doe.gov/fuelrenewable.html.

Source: U.S. Energy Information Administration Annual Energy Review 2009.

Wind[h]	Consumption							Total Renewable Energy	
	Biomass								
	Wood[i]		Waste[j]		Biofuels[k]		Total		
33,581	2,370,991		550,602	[R]	187,375	[R]	3,108,968	[R]	7,178,202	[R]
30,853	2,184,160		542,295		205,137	[R]	2,931,592	[R]	6,657,589	[R]
45,894	2,214,167		540,156		213,232	[R]	2,967,555	[R]	6,680,737	[R]
57,057	2,261,715		510,800	[R]	240,523		3,013,038	[R]	6,264,394	[R]
69,617	2,005,833		363,874		257,769	[R]	2,627,476	[R]	5,315,670	[R]
105,334	1,995,283		402,006		309,456	[R]	2,706,745	[R]	5,893,795	[R]
114,571	2,002,040		401,347		413,217	[R]	2,816,604	[R]	6,149,881	[R]
141,749	2,121,251	[R]	389,044	[R]	512,571	[R]	3,022,866	[R]	6,260,276	[R]
178,088	2,136,351	[R]	403,219	[R]	593,576	[R]	3,133,146	[R]	6,422,883	[R]
263,738	2,151,731	[R]	414,226	[R]	794,656	[R]	3,360,613	[R]	6,908,484	[R]
340,503	[R]	2,142,417	[R]	430,095	[R]	1,024,858	[R]	3,597,370	[R]	6,813,935	[R]
514,224	2,040,616		430,554		1,413,082		3,884,252		7,300,048	

TABLE 3 Energy Consumption by Sector, 1949–2008 (Billion Btu)

	End-use Sectors					
	Residential		Commercial[a]		Industrial[b]	
Year	Primary	Total[e]	Primary	Total[e]	Primary	
1949	4,475,121	5,613,938	2,660,963	3,660,910	12,626,532	
1950	4,847,590	6,006,806	2,824,267	3,883,472	13,881,079	
1951	5,124,031	6,399,747	2,727,158	3,862,700	15,118,070	
1952	5,178,644	6,580,694	2,661,902	3,862,377	14,661,778	
1953	5,074,890	6,581,124	2,500,330	3,758,937	15,328,413	
1954	5,286,016	6,869,767	2,444,814	3,720,157	14,305,657	
1955	5,633,095	7,303,271	2,547,641	3,881,530	16,090,702	
1956	5,866,467	7,689,809	2,592,274	4,008,279	16,562,350	
1957	5,771,579	7,739,679	2,434,391	3,945,887	16,512,867	
1958	6,155,096	8,230,400	2,541,202	4,103,153	15,797,985	
1959	6,223,822	8,447,378	2,630,274	4,353,069	16,518,951	
1960	6,688,963	9,077,668	2,702,042	4,588,973	16,977,066	
1961	6,814,611	9,325,376	2,743,974	4,706,925	16,993,115	
1962	7,122,112	9,825,201	2,901,109	5,013,919	17,589,807	
1963	7,135,126	10,034,384	2,896,921	5,226,862	18,365,964	
1964	7,161,257	10,290,804	2,949,284	5,438,649	19,426,503	
1965	7,328,128	10,688,770	3,150,462	5,819,530	20,123,911	
1966	7,549,262	11,218,183	3,383,741	6,299,383	21,029,715	
1967	7,740,902	11,669,926	3,738,448	6,870,845	21,012,628	
1968	7,963,327	12,368,421	3,866,000	7,296,778	21,872,069	
1969	8,276,760	13,205,347	4,045,666	7,795,301	22,653,721	
1970	8,352,750	13,798,057	4,196,051	8,307,155	22,974,833	
1971	8,456,799	14,277,629	4,282,718	8,681,492	22,732,356	
1972	8,655,327	14,890,531	4,369,078	9,144,775	23,532,489	
1973	8,250,226 [R]	14,929,771 [R]	4,381,061	9,506,982	24,740,862 [R]	
1974	7,927,553 [R]	14,683,314 [R]	4,221,192	9,362,537	23,816,329 [R]	
1975	8,005,740	14,841,755 [R]	4,022,853 [R]	9,465,906 [R]	21,454,213 [R]	
1976	8,408,252 [R]	15,440,661 [R]	4,332,587 [R]	10,035,225 [R]	22,685,371	
1977	8,207,376 [R]	15,688,729 [R]	4,217,258 [R]	10,177,267 [R]	23,192,694 [R]	
1978	8,272,389 [R]	16,155,929 [R]	4,268,843 [R]	10,480,604 [R]	23,276,491 [R]	
1979	7,933,806 [R]	15,841,970 [R]	4,333,251	10,626,851	24,211,500 [R]	
1980	7,453,254 [R]	15,786,781 [R]	4,074,270	10,562,769 [R]	22,610,288 [R]	
1981	7,057,589 [R]	15,295,246 [R]	3,805,343	10,601,863 [R]	21,338,216 [R]	
1982	7,154,067 [R]	15,557,340 [R]	3,835,241 [R]	10,847,354 [R]	19,075,786 [R]	
1983	6,840,628 [R]	15,456,669 [R]	3,806,026 [R]	10,922,977 [R]	18,578,019 [R]	
1984	7,220,681 [R]	15,998,041 [R]	3,968,567 [R]	11,436,092 [R]	20,197,515 [R]	
1985	7,160,776 [R]	16,088,348 [R]	3,694,740 [R]	11,443,724 [R]	19,467,805 [R]	
1986	6,921,722 [R]	16,029,197 [R]	3,656,730 [R]	11,603,742 [R]	19,098,662 [R]	
1987	6,940,917	16,321,196	3,736,106 [R]	11,943,383 [R]	19,977,070 [R]	
1988	7,372,024 [R]	17,186,278 [R]	3,957,548 [R]	12,575,483 [R]	20,884,381 [R]	
1989	7,586,093 [R]	17,858,128 [R]	4,004,442 [R]	13,202,580	20,897,403 [R]	
1990	6,570,463 [R]	17,014,681 [R]	3,858,007	13,332,926	21,208,225 [R]	
1991	6,758,442 [R]	17,490,321 [R]	3,905,836 [R]	13,512,501 [R]	20,854,317 [R]	
1992	6,963,482 [R]	17,426,920 [R]	3,951,199	13,453,951	21,786,666 [R]	
1993	7,155,529 [R]	18,288,984	3,933,859 [R]	13,835,823 [R]	21,784,999 [R]	
1994	6,990,569	18,181,216	3,978,979 [R]	14,111,283 [R]	22,422,272 [R]	
1995	6,946,268	18,577,978 [R]	4,063,119 [R]	14,697,525 [R]	22,747,660 [R]	
1996	7,471,455 [R]	19,562,439 [R]	4,234,533 [R]	15,181,207 [R]	23,443,770 [R]	

Energy Data / 179

	End-use Sectors			Electric Power Sector[c,d]		Balancing Item[f]		Total[g]	
	Industrial[b] Total[e]	Transportation							
		Primary	Total[e]	Primary					
	14,716,733	7,879,581	7,990,087	4,339,470		−165		31,981,503	
	16,232,875	8,383,528	8,492,594	4,679,283		21		34,615,768	
	17,669,234	8,933,753	9,042,162	5,070,830		188		36,974,030	
	17,301,575	8,907,235	9,003,096	5,338,183		82		36,747,825	
	18,200,961	9,030,518	9,123,484	5,730,355		−39		37,664,468	
	17,146,242	8,823,059	8,903,125	5,779,745		91		36,639,382	
	19,472,329	9,475,032	9,550,811	6,461,471		30		40,207,971	
	20,196,256	9,791,039	9,860,083	6,942,296		174		41,754,252	
	20,204,730	9,837,442	9,897,017	7,231,035		−128		41,787,186	
	19,306,571	9,952,797	10,004,893	7,197,936		11		41,645,028	
	20,315,979	10,298,441	10,349,357	7,794,295		−61		43,465,722	
	20,823,424	10,560,452	10,596,801	8,158,344		3		45,086,870	
	20,936,742	10,734,679	10,770,077	8,452,741		−103		45,739,017	
	21,768,109	11,185,922	11,220,519	9,028,798		−42		47,827,707	
	22,729,891	11,621,165	11,654,898	9,626,860		124		49,646,160	
	24,089,579	11,964,508	11,998,284	10,315,765		−140		51,817,177	
	25,074,894	12,400,149	12,433,906	11,014,449		121		54,017,221	
	26,397,297	13,069,166	13,101,884	11,984,863		−203		57,016,544	
	26,615,564	13,718,214	13,752,106	12,698,249		−333		58,908,107	
	27,888,371	14,831,020	14,865,583	13,886,738		238		62,419,392	
	29,114,339	15,470,880	15,506,152	15,174,112		−260		65,620,879	
	29,641,226	16,061,232	16,097,603	16,259,175		119		67,844,161	
	29,600,938	16,693,481	16,729,212	17,123,917		−307		69,288,965	
	30,952,764	17,681,086	17,716,273	18,466,362		−75		72,704,267	
	32,652,616 [R]	18,576,065		18,611,660	19,752,816	7,334 [R]		75,708,364	
	31,818,721 [R]	18,085,915 [R]	18,119,206 [R]	19,932,789		7,102 [R]		73,990,880	
	29,447,184 [R]	18,209,133 [R]	18,243,706 [R]	20,306,611		640 [R]		71,999,191	
	31,429,542	19,065,144 [R]	19,099,331 [R]	21,513,405		7,613 [R]		76,012,373	
	32,306,559 [R]	19,784,143 [R]	19,819,581 [R]	22,590,665		7,418 [R]		77,999,554	
	32,733,452 [R]	20,580,415 [R]	20,614,766 [R]	23,586,613		1,619 [R]		79,986,371	
	33,962,118 [R]	20,436,369 [R]	20,470,711 [R]	23,986,723		1,564		80,903,214	
	32,077,090 [R]	19,658,353 [R]	19,696,034 [R]	24,326,509		−1,080 [R]		78,121,594	
	30,756,076 [R]	19,476,200 [R]	19,512,537 [R]	24,488,373		2,766 [R]		76,168,488	[R]
	27,656,788 [R]	19,050,580 [R]	19,087,723 [R]	24,033,531		4,189 [R]		73,153,394	[R]
	27,481,484 [R]	19,132,451 [R]	19,175,075 [R]	24,679,081		2,796 [R]		73,039,001	[R]
	29,624,598 [R]	19,606,799 [R]	19,653,933 [R]	25,719,102		2,794 [R]		76,715,459	[R]
	28,877,080 [R]	20,040,687 [R]	20,087,315 [R]	26,132,459		−3,903 [R]		76,492,565	[R]
	28,333,363 [R]	20,739,703 [R]	20,788,771 [R]	26,338,257		3,452		76,758,526	[R]
	29,443,635 [R]	21,419,125 [R]	21,469,449 [R]	27,104,445		−2,533 [R]		79,175,130	[R]
	30,738,557 [R]	22,266,855 [R]	22,318,176 [R]	28,337,687		3,364 [R]		82,821,858	[R]
	31,397,833 [R]	22,424,597 [R]	22,478,708 [R]	30,024,713 [4]		8,999 [R]		84,946,248	[R]
	31,895,492 [R]	22,366,185 [R]	22,419,888 [R]	30,660,106		−9,335 [R]		84,653,651	[R]
	31,486,967 [R]	22,065,034 [R]	22,118,484 [R]	31,024,645		595 [R]		84,608,869	[R]
	32,661,236 [R]	22,363,309 [R]	22,415,918 [R]	30,893,368		355 [R]		85,958,380	[R]
	32,721,292 [R]	22,716,447 [R]	22,769,843 [R]	32,025,108		−10,490 [R]		87,605,453	[R]
	33,607,366 [R]	23,311,806 [R]	23,367,224 [R]	32,563,463		−5,698		89,261,391	[R]
	34,046,786 [R]	23,793,148 [R]	23,848,651 [R]	33,620,747		3,148 [R]		91,174,089	[R]
	34,988,791 [R]	24,383,906 [R]	24,438,890 [R]	34,637,665		4,336 [R]		94,175,664	[R]

TABLE 3 (*Continued*)

Year	End-use Sectors					
	Residential		Commercial[a]		Industrial[b]	
	Primary	Total[e]	Primary	Total[e]	Primary	
1997	7,039,505 [R]	19,025,680 [R]	4,256,507 [R]	15,693,953 [R]	23,721,864 [R]	
1998	6,423,825 [R]	19,020,712 [R]	3,963,729 [R]	15,979,296 [R]	23,210,838 [R]	
1999	6,783,779 [R]	19,620,860 [R]	4,007,378 [R]	16,383,617 [R]	22,990,578 [R]	
2000	7,168,979 [R]	20,487,621 [R]	4,227,143 [R]	17,176,087 [R]	22,870,804 [R]	
2001	6,878,917 [R]	20,106,132 [R]	4,036,108 [R]	17,141,259 [R]	21,835,587 [R]	
2002	6,938,187 [R]	20,873,763 [R]	4,099,189 [R]	17,366,740	21,857,313 [R]	
2003	7,251,896	21,208,021 [R]	4,238,672 [R]	17,351,447 [R]	21,575,582 [R]	
2004	7,019,274 [R]	21,177,889 [R]	4,180,422 [R]	17,664,445 [R]	22,454,620 [R]	
2005	6,920,879 [R]	21,697,240 [R]	4,013,701 [R]	17,875,276 [R]	21,465,855 [R]	
2006	6,190,514 [R]	20,769,777 [R]	3,703,258 [R]	17,723,994 [R]	21,632,057 [R]	
2007	6,625,793 [R]	21,619,373 [R]	3,895,928 [R]	18,287,222 [R]	21,454,002 [R]	
2008 P	6,778,379	21,636,900	3,972,150	18,541,387	20,630,137	

[a]Commercial sector, including commercial combined-heat-and-power (CHP) and commercial electricity-only plants.

[b]Industrial sector, including industrial CHP and industrial electricity-only plants.

[c]Electricity-only and CHP plants within the NAICS 22 category whose primary business is to sell electricity, or electricity and heat, to the public.

[d]Through 1988, data are for electric utilities only; beginning in 1989, data are for electric utilities and independent power producers.

[e]Total energy consumption in the end-use sectors consists of primary energy consumption, electricity retail sales, and electrical system energy losses.

[f]A balancing item. The sum of primary consumption in the five energy-use sectors equals the sum of total consumption in the four end-use sectors. However, total energy consumption does not equal the sum of the sectoral components because of the use of sector-specific conversion factors for natural gas and coal.

[g]Primary energy consumption total.

R = Revised. P = Preliminary. (s) = Less than 0.5 trillion Btu.

Source: U.S. Energy Information Administration Annual Energy Review 2009.

End-use Sectors				Electric Power Sector[c,d]		Balancing Item[f]		Total[g]	
Industrial[b]		Transportation							
Total[e]		Primary		Total[e]		Primary			
35,288,218	[R]	24,697,145	[R]	24,751,817	[R]	35,044,648		6,142	[R]	94,765,811	[R]
34,928,190	[R]	25,203,168	[R]	25,258,473	[R]	36,385,110		−3,378	[R]	95,183,293	[R]
34,855,491	[R]	25,893,727	[R]	25,951,203	[R]	37,135,709		6,281	[R]	96,817,452	[R]
34,757,478	[R]	26,491,500	[R]	26,551,610	[R]	38,214,371		2,304	[R]	98,975,100	[R]
32,806,204	[R]	26,215,564	[R]	26,278,577	[R]	37,365,995		−6,084	[R]	96,326,089	[R]
32,764,483	[R]	26,787,738	[R]	26,848,508	[R]	38,171,067		4,820	[R]	97,858,314	[R]
32,649,843	[R]	26,927,646	[R]	27,002,137	[R]	38,217,654	[R]	−2,908	[R]	98,208,541	[R]
33,609,067	[R]	27,820,116	[R]	27,899,279	[R]	38,876,247	[R]	−55	[R]	100,350,624	[R]
32,545,253	[R]	28,279,693	[R]	28,361,295	[R]	39,798,935	[R]	5,694	[R]	100,484,758	[R]
32,541,235	[R]	28,761,209	[R]	28,840,577	[R]	39,588,544	[R]	−385	[R]	99,875,196	[R]
32,523,120	[R]	29,046,175	[R]	29,134,189	[R]	40,542,007	[R]	−10,049	[R]	101,553,855	[R]
31,210,299		27,842,133		27,924,560		40,090,347		−9,512		99,303,634	

TABLE 4 Household End Uses: Fuel Types and Appliances, Selected Years, 1978–2005

Appliance	1978	1979	1980	1981	1982	1984	1987
Total households (millions)	77	78	82	83	84	86	91
	\multicolumn{7}{c}{Percent of Households}						
Space heating—Main fuel							
Natural Gas	55	55	55	56	57	55	55
Electricity[a]	16	17	18	17	16	17	20
Liquefied petroleum gases	4	5	5	4	5	5	5
Distillate fuel oil	20	17	15	14	13	12	12
Wood	2	4	6	6	7	7	6
Other [b] or no space heating	3	2	2	3	3	3	3
Air conditioning—Equipment							
Central System[c]	23	24	27	27	28	30	34
Window/wall unit[c]	33	31	30	31	30	30	30
None	44	45	43	42	42	40	36
Water heating—Main fuel							
Natural Gas	55	55	54	55	56	54	54
Electricity[a]	33	33	32	33	32	33	35
Liquefied petroleum gases	4	4	4	4	4	4	3
Distillate fuel oil	8	7	9	7	7	6	6
Other[b] or no water heating	0	0	1	1	1	1	1
Appliances and electronics							
Refrigerator[d]	100	NA	100	100	100	100	100
One	86	NA	86	87	86	88	86
Two or more	14	NA	14	13	13	12	14
Separate freezer	35	NA	38	38	37	37	34
Clothes washer	74	NA	74	73	71	73	75
Clothes dryer—Total	59	NA	61	61	60	62	66
Natural gas	14	NA	14	16	15	16	15
Electric	45	NA	47	45	45	46	51
Dishwasher	35	NA	37	37	36	38	43
Range/stove/oven	99	NA	99	100	99	99	99
Natural gas	48	NA	46	46	47	46	43
Electric	53	NA	57	56	56	57	60
Microwave oven	8	NA	14	17	21	34	61
Television	NA	NA	98	98	98	98	98
One or two	NA	NA	85	84	83	80	75
Three or more	NA	NA	14	14	15	18	23
Personal computer	NA	NA	NA	NA	NA	NA	NA
One	NA	NA	NA	NA	NA	NA	NA
Two or more	NA	NA	NA	NA	NA	NA	NA

[a]Retail electricity.

[b]Kerosene, solar, or other fuel.

[c]Households with both a central system and a window or wall unit are counted only under "Central System."

[d]Fewer than 0.5 percent of the households do not have a refrigerator.

R = Revised. NA = Not available. (s) = Less than 0.5 percent.

Note: Data are estimates. For years not shown, there are no data available. For related information, see http://www.eia.doe.gov/emeu/recs.

Sources: For 1978 and 1979, Energy Information Administration (EIA), Form EIA-84, "Residential Energy Consumption Survey"; for 1980–2005, EIA, Form EIA-457, "Residential Energy Consumption Survey."

Energy Data / 183

	Year					Change
	1990	1993	1997	2001	2005	1980 to 2005
	94	97	101	107	111	29
	Percent of Households					
	55	53	52 [R]	55	52	-3
	23	26	29	29	30	12
	5	5	5	5	5	0
	11	11	9	7 [R]	7	-8
	4	3	2	2	3	-3
	2	2	2	2 [R]	3 [R]	1 [R]
	39	44	47	55	59	32
	29	25	25	23	25	-5
	32	32	28	23	16	-27
	53	53	52	54	53	-1
	37	38	39	38	39	7
	3	3	3	3	4	0
	5	5	5	4	4	-5
	1	1	1	0	0	-1
	100	100	100	100	100	0
	84	85	85	83	78	-8
	15	15	15	17	22	8
	34	35	33	32	32	-6
	76	77	77	79	83	9
	69	70	71	74	79	18
	16	14 [R]	15 [R]	16 [R]	17 [R]	3 [R]
	53	57	55	57	61	14
	45	45	50	53	58	21
	100	100	99	100	99	0
	42	33	35	35	35	-11
	59	63	62	62	62	5
	79	84	83	86	88	74
	99	99	99 [R]	99 [R]	99 [R]	1 [R]
	71	70	69	63	56	-29
	28	28	29	36	43	29
	NA [R]	NA [R]	35	56	68	NA
	NA	NA	29	42	45	NA
	NA	NA	6	15	23	NA

TABLE 5 World Primary Energy Consumption by Region, 1997–2006 (Quadrillion Btu)

Region and Country	1997		1998		1999		2000	
North America	**113.13**		**113.53**		**115.82**		**118.26**	
Canada	12.67	[R]	12.37	[R]	12.96	[R]	12.95	[R]
Mexico	5.68		5.96		6.04		6.32	
United States	94.77		95.18		96.82		98.98	
Other	.02		.02		.02		.02	
Central and South America	**19.45**		**20.12**		**20.27**		**20.84**	
Argentina	2.47		2.58		2.61		2.67	
Brazil	7.86		8.12	[R]	8.27	[R]	8.55	[R]
Venezuela	2.66		2.85		2.73		2.77	
Other	6.46		6.57		6.67		6.85	
Europe [a]	**79.87**	[R]	**80.44**	[R]	**80.51**	[R]	**81.53**	[R]
Belgium	2.65	[R]	2.70	[R]	2.66	[R]	2.73	[R]
France	10.36		10.58		10.71		10.85	
Germany	14.36		14.34		14.13		14.26	
Italy	7.22		7.43		7.56		7.63	
Netherlands	3.70		3.70		3.69		3.79	
Poland	4.09	[R]	3.85		3.98		3.62	
Spain	4.76		4.99		5.26		5.62	
Sweden	2.32	[R]	2.40	[R]	2.37	[R]	2.27	[R]
Turkey	2.93		3.00		2.91		3.16	
United Kingdom	9.75		9.74	[R]	9.79	[R]	9.72	[R]
Other	17.74		17.72		17.47	[R]	17.87	
Eurasia [b]	**39.02**	[R]	**38.73**	[R]	**39.83**	[R]	**40.61**	[R]
Russia	25.81	[R]	25.93	[R]	27.01	[R]	27.47	[R]
Ukraine	6.07		5.85		5.76		5.75	
Uzbekistan	1.88		1.84		1.86		1.94	
Other	5.26		5.11		5.19		5.45	[R]
Middle East	**15.61**		**16.28**	[R]	**16.62**		**17.32**	
Iran	4.43		4.58		4.83		5.01	
Saudi Arabia	4.37		4.54		4.60		4.85	
Other	6.81		7.15		7.18		7.46	
Africa	**11.40**		**11.30**	[R]	**11.62**	[R]	**12.03**	[R]
Egypt	1.79		1.85		1.92		2.00	
South Africa	4.56		4.35		4.46		4.59	
Other	5.05		5.10	[R]	5.23		5.44	[R]
Asia and Oceania [a]	**102.89**	[R]	**101.98**	[R]	**105.28**	[R]	**107.33**	[R]
Australia	4.56		4.59		4.82		4.85	
China	37.91		37.32		37.23		37.18	[R]
India	11.64		12.17		12.99		13.46	
Indonesia	3.66		3.56		3.91		4.06	
Japan	21.91	[R]	21.52	[R]	21.97	[R]	22.43	[R]
Malaysia	1.67		1.69		1.74		1.87	

	2001		2002		2003		2004		2005		2006 P
	115.36		**117.25**		**118.20**		**120.74**		**121.62**		**121.18**
	12.76	[R]	13.13	[R]	13.56	[R]	13.84	[R]	14.23	[R]	13.95
	6.26		6.25	[R]	6.42	[R]	6.53		6.86	[R]	7.36
	96.33		97.86		98.21		100.35		100.51	[R]	99.86
	.02		.02		.02		.02		.02		.02
	21.16		**21.12**		**21.61**		**22.44**		**23.40**		**24.18**
	2.61		2.48	[R]	2.67		2.78		2.95	[R]	3.15
	8.47	[R]	8.58	[R]	8.69	[R]	9.02	[R]	9.37	[R]	9.64
	3.03		2.93		2.72		2.93		3.12	[R]	3.19
	7.05		7.13		7.54		7.71		7.96		8.20
	82.77	[R]	**82.50**	[R]	**84.24**	[R]	**85.70**	[R]	**86.18**	[R]	**86.42**
	2.70	[R]	2.68	[R]	2.78	[R]	2.81	[R]	2.78	[R]	2.75
	11.08		11.00		11.11	[R]	11.39		11.36	[R]	11.44
	14.62		14.33		14.59	[R]	14.74	[R]	14.50	[R]	14.63
	7.67	[R]	7.70		7.99	[R]	8.08	[R]	8.14	[R]	8.07
	3.93		3.94		4.00		4.11		4.23	[R]	4.14
	3.45		3.44		3.60		3.70		3.68	[R]	3.86
	5.87		5.95		6.26		6.39	[R]	6.51	[R]	6.51
	2.40	[R]	2.27	[R]	2.17	[R]	2.30	[R]	2.33	[R]	2.22
	2.89		3.15		3.32		3.51		3.73	[R]	3.91
	9.86	[R]	9.72	[R]	9.86	[R]	9.88	[R]	9.92	[R]	9.80
	18.28	[R]	18.33	[R]	18.56	[R]	18.77	[R]	19.01	[R]	19.10
	40.94	[R]	**41.59**	[R]	**43.37**	[R]	**44.69**	[R]	**45.79**	[R]	**45.88**
	27.72	[R]	27.93	[R]	28.77	[R]	29.60	[R]	30.06	[R]	30.39
	5.64		5.82		6.28		6.26		6.32	[R]	5.87
	2.03		2.08		2.10		2.22		2.13	[R]	2.21
	5.55	[R]	5.75	[R]	6.22	[R]	6.62	[R]	7.27	[R]	7.41
	17.95		**18.98**		**19.76**		**20.89**		**22.75**	[R]	**23.81**
	5.39		5.89		6.18		6.39		7.22	[R]	7.69
	5.14		5.38		5.76		6.21		6.59	[R]	6.89
	7.42		7.71		7.82		8.29		8.93	[R]	9.23
	12.63	[R]	**12.72**		**13.36**	[R]	**13.97**	[R]	**14.54**	[R]	**14.50**
	2.23	[R]	2.26	[R]	2.44	[R]	2.59		2.73	[R]	2.54
	4.66		4.54		4.88		5.21		5.12	[R]	5.18
	5.74	[R]	5.91	[R]	6.04	[R]	6.18	[R]	6.69	[R]	6.77
	111.34	[R]	**116.41**	[R]	**125.48**	[R]	**138.71**	[R]	**147.78**	[R]	**156.31**
	5.02		5.13		5.14	[R]	5.26	[R]	5.57	[R]	5.61
	39.44	[R]	43.30	[R]	50.62	[R]	59.99	[R]	66.80	[R]	73.81
	13.94		13.84		14.29		15.54	[R]	16.34	[R]	17.68
	4.46		4.64		4.56	[R]	4.88	[R]	4.91	[R]	4.15
	22.24	[R]	22.15	[R]	22.15	[R]	22.74	[R]	22.74	[R]	22.79
	2.11		2.18		2.42		2.66		2.58	[R]	2.56

TABLE 5 (*Continued*)

Region and Country	1997		1998		1999		2000		
South Korea	7.41		6.83		7.55		7.89		
Taiwan	3.21		3.40		3.55		3.77		
Thailand	2.60		2.44		2.50		2.58		
Other	8.34	[R]	8.47	[R]	9.01	[R]	9.23	[R]	
World	381.35	[R]	382.38	[R]	389.95	[R]	397.93	[R]	

[a]Excludes countries that were part of the former USSR.

[b]Includes only countries that were part of the former USSR.

R = Revised. P = Preliminary.

Notes: Data in this table do not include recent updates for the United States or for other countries (see http://tonto.eia.doe.gov/cfapps/ipdbproject/IEDIndex3.cfm). World primary energy consumption includes consumption of petroleum products (including natural-gas plant liquids and crude oil burned as fuel), dry natural gas, and coal (including net imports of coal coke) and the consumption of net electricity generated from nuclear electric power, hydroelectric power, wood, waste, geothermal, solar, and wind. It also includes, for the United States, the consumption of renewable energy by the end-use sectors. Totals may not equal sum of components due to independent rounding. For related information, see http://www.eia.doe.gov/international.

Source: Energy Information Administration, "International Energy Annual 2006" (June–December 2008), Table E1.

	2001		2002		2003		2004		2005		2006 P
	8.10		8.39	[R]	8.64	[R]	8.91	[R]	9.23	[R]	9.45
	3.86		4.02		4.21		4.36		4.43	[R]	4.57
	2.70		2.94		3.22		3.45	[R]	3.67	[R]	3.74
	9.47	[R]	9.80	[R]	10.23	[R]	10.92	[R]	11.52	[R]	11.97
	402.15	[R]	410.56	[R]	426.02	[R]	447.15	[R]	462.06	[R]	472.27

TABLE 6 World Crude Oil and Natural Gas Reserves, January 1, 2008

Region and Country	Crude Oil		Natural Gas	
	Oil & Gas Journal	World Oil	Oil & Gas Journal	World Oil
	Billion Barrels		Trillion Cubic Feet	
North America	211.6	57.5	309.8	314.1
Canada	178.6[a]	25.2[b]	58.2	58.3
Mexico	11.7	11.1	13.9	18.1
United States	21.3	21.3	237.7	237.7
Central and South America	109.9	104.8	261.8	247.0
Argentina	2.6	2.7	15.8	16.5
Bolivia	.5	.5	26.5	28.0
Brazil	12.2	12.5	12.3	12.9
Chile	.2	.0	3.5	1.0
Colombia	1.5	1.5	4.3	6.7
Cuba	.1	.7	2.5	.8
Ecuador	4.5	4.8	NA	.3
Peru	.4	.4	11.9	12.0
Trinidad and Tobago	.7	.6	18.8	16.7
Venezuela	87.0	81.0	166.3	152.0
Other[c]	.2	.2	(s)	(s)
Europe[d]	14.3	13.8	172.0	169.0
Austria	.1	.1	.6	1.1
Croatia	.1	.1	1.0	1.1
Denmark	1.2	1.1	2.5	2.6
Germany	.4	.2	9.0	5.2
Hungary	(s)	.1	.3	.6
Italy	.4	.4	3.3	3.0
Netherlands	.1	.2	50.0	48.8
Norway	6.9	6.7	79.1	81.7
Poland	.1	.2	5.8	4.7
Romania	.6	.5	2.2	4.2
Serbia	.1	NR	1.7	NR
United Kingdom	3.6	3.6	14.6	14.0
Other[c]	.8	.7	1.9	2.1
Eurasia[e]	98.9	126.0	2,014.8	2,104.0
Azerbaijan	7.0	NR	30.0	NR
Kazakhstan	30.0	NR	100.0	NR
Russia	60.0	76.0	1,680.0	1,654.0
Turkmenistan	.6	NR	100.0	NR
Ukraine	.4	NR	39.0	NR
Uzbekistan	.6	NR	65.0	NR
Other[c]	.3	50.0	.8	450.0

	Crude Oil		Natural Gas	
	Oil & Gas Journal	World Oil	Oil & Gas Journal	World Oil
Region and Country	Billion Barrels		Trillion Cubic Feet	
Middle East	748.3	727.3	2,548.9	2,570.2
Bahrain	.1	NR	3.3	NR
Iran	138.4	137.0	948.2	985.0
Iraq	115.0	126.0	111.9	91.0
Kuwait[f]	104.0	99.4	56.0	66.3
Oman	5.5	5.7	30.0	32.0
Qatar	15.2	20.0	905.3	903.2
Saudi Arabia[f]	266.8	264.8	253.1	254.0
Syria	2.5	2.9	8.5	12.1
United Arab Emirates	97.8	68.1	214.4	196.3
Yemen	3.0	2.7	16.9	16.8
Other[c]	(s)	.7	1.3	13.6
Africa	114.8	114.7	489.6	504.2
Algeria	12.2	11.9	159.0	160.0
Angola	9.0	9.5	9.5	5.7
Cameroon	.2	NR	4.8	NR
Congo (Brazzaville)	1.6	1.9	3.2	4.1
Egypt	3.7	3.7	58.5	68.5
Equatorial Guinea	1.1	1.7	1.3	3.4
Gabon	2.0	3.2	1.0	2.5
Libya	41.5	36.5	50.1	52.8
Mozambique	.0	.0	4.5	.0
Nigeria	36.2	37.2	184.0	184.5
Sudan	5.0	6.7	3.0	4.0
Tunisia	.4	.6	2.3	3.5
Other[c]	1.9	1.8	7.6	15.4
Asia and Oceania[d]	34.3	40.0	415.4	527.6
Australia	1.5	4.2	30.0	151.9
Bangladesh	(s)	NR	5.0	NR
Brunei	1.1	1.1	13.8	11.0
Burma	.1	.2	10.0	15.0
China	16.0	18.1	80.0	61.8
India	5.6	4.0	38.0	31.8
Indonesia	4.4	4.5	93.9	92.0
Japan	(s)	NR	.7	NR
Malaysia	4.0	5.5	83.0	88.0
New Zealand	.1	.1	1.0	2.0
Pakistan	.3	.3	28.0	29.8
Papua New Guinea	.1	.2	8.0	14.7
Thailand	.5	.4	11.7	11.2
Vietnam	.6	1.3	6.8	8.2
Other[c]	.2	.2	5.5	10.2
World	1,332.0	1,184.2	6,212.3	6,436.0

[a] Comprises 5.4 billion barrels of conventional crude oil and condensate and 173.2 billion barrels of bitumen in Alberta's oil sands.

[b] *World Oil* states the following about its Canadian crude oil reserves estimate: "*conventional* crude reserves are 4.9 Bbbl [billion barrels]. Alberta's estimates of *established* oil sands reserves of 174 Bbbl are not proved; that would require at least 350 Tcf [trillion cubic feet] of gas delivered to northern Alberta, and/or implementation of future technologies. *Oil sands* reserve estimate is based on 50 years times current production capacity."

[c] Includes data for those countries not separately reported.

[d] Excludes countries that were part of the former USSR.

[e] Includes only countries that were part of the former USSR.

[f] Data for Kuwait and Saudi Arabia include one-half of the reserves in the neutral zone between Kuwait and Saudi Arabia.

NA = Not available. NR = Not separately reported. (s) = Less than 0.05 billion barrels.

Notes: All reserve figures are proved reserves, except as noted. Totals may not equal sum of components as a result of independent rounding. For related information, see http://www.eia.doe.gov/international.

Sources: U.S. data, Energy Information Administration, *U.S. Crude Oil, Natural Gas, and Natural Gas Liquids Reserves, 2007 Annual Report;* All other data, PennWell Corporation, *Oil & Gas Journal* 105, no. 48 (December 24, 2007) and Gulf Publishing Company, *World Oil* 229, no. 9 (September 2008).

TABLE 7 World Recoverable Reserves of Coal, 2005 (Million Short Tons)

Region and Country	Anthracite and Bituminous Coal		Subbituminous Coal and Lignite		Total	
North America	126,271	[R]	145,206	[R]	271,477	[R]
Canada	3,826		3,425		7,251	
Greenland	0		202		202	
Mexico	948		387		1,335	
United States[a]	121,496	[R]	141,193	[R]	262,689	[R]
Central and South America	7,969		9,973		17,941	
Brazil	0		7,791		7,791	
Chile	34		1,268		1,302	
Colombia	7,251		420		7,671	
Peru	154		0		154	
Other	529		494		1,023	
Europe[b]	9,296		41,485		50,781	
Bulgaria	6		2,195		2,200	
Czech Republic	1,844		3,117		4,962	
Former Serbia and Montenegro	7		15,299		15,306	
Germany	168		7,227		7,394	
Greece	0		4,299		4,299	
Hungary	219		3,420		3,640	
Poland	6,627		1,642		8,270	
Romania	13		452		465	
Turkey	0		2,000		2,000	
United Kingdom	171		0		171	
Other	241		1,834		2,076	

TABLE 7 (*Continued*)

Region and Country	Anthracite and Bituminous Coal	Subbituminous Coal and Lignite	Total	
Eurasia[c]	103,186	145,931	249,117	
Kazakhstan	31,052	3,450	34,502	
Russia	54,110	118,964	173,074	
Ukraine	16,922	20,417	37,339	
Uzbekistan	1,102	2,205	3,307	
Other	0	895	895	
Middle East	1,528	0	1,528	
Iran	1,528	0	1,528	
Africa	54,488	192	54,680	
Botswana	44	0	44	
South Africa	52,911	0	52,911	
Zimbabwe	553	0	553	
Other	980	192	1,172	
Asia and Oceania[b]	169,994	113,813	283,807	
Australia	40,896	43,541	84,437	
China	68,564	57,651	126,215	
India	57,585	4,694	62,278	
Indonesia	1,897	2,874	4,771	
North Korea	331	331	661	
Pakistan	1	2,184	2,185	
Thailand	0	1,493	1,493	
Other	721	1,046	1,767	
World	472,731 [R]	456,599 [R]	929,331	[R]

[a]U.S. data are as of the end of 2007, 2 years later than the other data on this table.

[b]Excludes countries that were part of the former USSR.

[c]Includes only countries that were part of the former USSR.

R = Revised.

Notes: Data are at end of year. World Energy Council data represent "proved recoverable reserves," which are the tonnage within the "proved amount in place" that can be recovered (extracted from the earth in raw form) under present and expected local economic conditions with existing, available technology. The Energy Information Administration does not certify the international reserves data but reproduces the information as a matter of convenience for the reader. U.S. reserves represent estimated recoverable reserves from the Demonstrated Reserve Base, which includes both measured and indicated tonnage. The U.S. term "measured" approximates the term "proved" as used by the World Energy Council. The U.S. "measured and indicated" data have been combined and cannot be recaptured as "measured alone." Totals may not equal sum of components as a result of independent rounding. For related information, see http://www.eia.doe.gov/international.

Sources: U.S. data based on EIA, *Annual Coal Report 2007,* Table 15, and unpublished file data of the Coal Reserves Data Base; All other data, World Energy Council, *2007 Survey of Energy Resources.*

TABLE 8 World Carbon Dioxide Emissions from Energy Consumption, 1997–2006 (Million Metric Tons of Carbon Dioxide)[a]

Region and Country	1997		1998		1999		2000	
North America	6,492	[R]	6,547	[R]	6,615	[R]	6,810	[R]
Canada	549	[R]	554	[R]	568	[R]	565	[R]
Mexico	350	[R]	372	[R]	364	[R]	383	[R]
United States	5,592	[R]	5,620	[R]	5,682	[R]	5,860	[R]
Other	1		1		1		1	
Central and South America	950	[R]	975	[R]	984	[R]	993	[R]
Argentina	130		136	[R]	140	[R]	138	[R]
Brazil	326	[R]	325	[R]	336	[R]	345	[R]
Venezuela	135	[R]	142		133		134	
Other	359	[R]	372	[R]	374	[R]	375	
Europe[b]	4,503	[R]	4,487	[R]	4,436	[R]	4,500	[R]
Belgium	146	[R]	151	[R]	143	[R]	149	[R]
France	385	[R]	410	[R]	404	[R]	402	[R]
Germany	889	[R]	872	[R]	841	[R]	857	[R]
Italy	425	[R]	441	[R]	441	[R]	448	[R]
Netherlands	240	[R]	242	[R]	239	[R]	252	[R]
Poland	339	[R]	316	[R]	329	[R]	295	[R]
Romania	120	[R]	101	[R]	91		93	
Spain	272	[R]	282	[R]	309	[R]	327	[R]
Turkey	182	[R]	184	[R]	182	[R]	202	[R]
United Kingdom	569	[R]	564	[R]	559	[R]	561	[R]
Other	935	[R]	924	[R]	898	[R]	913	[R]
Eurasia[c]	2,244	[R]	2,235	[R]	2,320	[R]	2,356	[R]
Kazakhstan	120	[R]	116	[R]	133	[R]	143	[R]
Russia	1,483	[R]	1,482	[R]	1,560	[R]	1,582	[R]
Ukraine	344	[R]	333	[R]	328	[R]	327	[R]
Uzbekistan	103		102	[R]	103		106	
Other	194	[R]	201	[R]	195	[R]	197	[R]
Middle East	989	[R]	1,019	[R]	1,057	[R]	1,094	[R]
Iran	291	[R]	295	[R]	317	[R]	321	[R]
Saudi Arabia	255	[R]	258	[R]	264	[R]	291	[R]
Other	443	[R]	467	[R]	475	[R]	483	[R]
Africa	872	[R]	861	[R]	877	[R]	892	[R]
Egypt	112	[R]	115	[R]	117	[R]	119	
South Africa	388	[R]	370	[R]	381	[R]	392	[R]
Other	371	[R]	376	[R]	378	[R]	381	[R]
Asia and Oceania[c]	7,197	[R]	7,035	[R]	7,247	[R]	7,366	[R]
Australia	334	[R]	340	[R]	359	[R]	360	[R]
China	3,133	[R]	3,029	[R]	2,992	[R]	2,967	[R]
India	878	[R]	914	[R]	971	[R]	1,012	[R]
Indonesia	247	[R]	241	[R]	266	[R]	274	[R]
Japan	1,161	[R]	1,116	[R]	1,158	[R]	1,204	[R]

	2001		2002		2003		2004		2005		2006 P
	6,697	[R]	6,782	[R]	6,870	[R]	6,970	[R]	7,034	[R]	6,954
	554	[R]	573	[R]	602	[R]	615	[R]	632	[R]	614
	380		384		389		385	[R]	407	[R]	436
	5,762	[R]	5,824	[R]	5,878	[R]	5,969	[R]	5,994	[R]	5,903
	1		1		1		1		1		1
	1,016	[R]	1,005	[R]	1,023	[R]	1,066	[R]	1,111	[R]	1,138
	128	[R]	121	[R]	134	[R]	141	[R]	152	[R]	162
	349	[R]	347	[R]	346	[R]	356	[R]	371	[R]	377
	149	[R]	147	[R]	134	[R]	143	[R]	150	[R]	152
	389		390	[R]	408	[R]	426	[R]	438		447
	4,559	[R]	4,532	[R]	4,679	[R]	4,713	[R]	4,717	[R]	4,721
	146	[R]	143	[R]	151	[R]	154	[R]	151	[R]	148
	406	[R]	402	[R]	409	[R]	416	[R]	414	[R]	418
	878	[R]	857	[R]	874	[R]	872	[R]	853	[R]	858
	445	[R]	453	[R]	475	[R]	470	[R]	473	[R]	468
	278	[R]	259	[R]	261	[R]	271	[R]	273	[R]	260
	279	[R]	276	[R]	289	[R]	295	[R]	290	[R]	303
	102	[R]	100	[R]	100	[R]	100	[R]	98	[R]	99
	332	[R]	349	[R]	357	[R]	371	[R]	384	[R]	373
	184	[R]	195	[R]	207	[R]	211	[R]	231	[R]	236
	575	[R]	564	[R]	575	[R]	582	[R]	585	[R]	586
	934	[R]	934	[R]	980	[R]	972	[R]	966	[R]	973
	2,332	[R]	2,354	[R]	2,471	[R]	2,529	[R]	2,600	[R]	2,601
	148	[R]	154	[R]	166	[R]	185	[R]	203	[R]	213
	1,571	[R]	1,572	[R]	1,627		1,663	[R]	1,699	[R]	1,704
	319	[R]	327	[R]	357	[R]	347	[R]	350	[R]	329
	111		114	[R]	115	[R]	122	[R]	117	[R]	121
	184	[R]	188		206	[R]	212	[R]	231	[R]	233
	1,119	[R]	1,175	[R]	1,240	[R]	1,330	[R]	1,444	[R]	1,505
	334	[R]	365	[R]	387	[R]	407	[R]	446	[R]	471
	301	[R]	312	[R]	347	[R]	389	[R]	406	[R]	424
	483	[R]	499	[R]	506	[R]	535	[R]	593	[R]	610
	923	[R]	924	[R]	975	[R]	1,025	[R]	1,062	[R]	1,057
	130	[R]	134	[R]	144	[R]	153	[R]	161	[R]	152
	399	[R]	385	[R]	418	[R]	448	[R]	438	[R]	444
	394	[R]	405	[R]	413	[R]	424	[R]	463	[R]	461
	7,608	[R]	8,050	[R]	8,806	[R]	9,821	[R]	10,517	[R]	11,220
	374	[R]	383	[R]	381	[R]	391	[R]	417	[R]	417
	3,108	[R]	3,441	[R]	4,062	[R]	4,847	[R]	5,429	[R]	6,018
	1,035	[R]	1,034	[R]	1,048	[R]	1,151	[R]	1,194	[R]	1,293
	300	[R]	315	[R]	305	[R]	323	[R]	324	[R]	280
	1,197	[R]	1,203	[R]	1,253	[R]	1,258	[R]	1,250	[R]	1,247

TABLE 8 (Continued)

Region and Country	1997		1998		1999		2000	
Malaysia	102		103	[R]	107	[R]	112	[R]
South Korea	435	[R]	375	[R]	433	[R]	446	[R]
Taiwan	210	[R]	225	[R]	224	[R]	252	[R]
Thailand	177	[R]	162	[R]	171	[R]	162	[R]
Other	520	[R]	530	[R]	567	[R]	578	[R]
World	23,247	[R]	23,160	[R]	23,535	[R]	24,011	[R]

[a]Metric tons of carbon dioxide can be converted to metric tons of carbon equivalent by multiplying by 12/44.

[b]Excludes countries that were part of the former USSR.

[c]Includes only countries that were part of the former USSR.

R = Revised. P = Preliminary.

Notes: Data in this table do not include recent updates (see http://tonto.eia.doe.gov/cfapps/ipdbproject/IEDIndex3.cfm). Data include carbon dioxide emissions from fossil-fuel energy consumption and natural-gas flaring. Totals may not equal sum of components as a result of independent rounding. For related information, see http://www.eia.doe.gov/international.

Source: Energy Information Administration, "International Energy Annual 2006" (June–December 2008), Table H.1co2.

2001		2002		2003		2004		2005		2006 P
125	[R]	134	[R]	150	[R]	166	[R]	160	[R]	164
452	[R]	468	[R]	478	[R]	489	[R]	497	[R]	515
249	[R]	274	[R]	290	[R]	287	[R]	290	[R]	300
172	[R]	187	[R]	206	[R]	226	[R]	243	[R]	245
594	[R]	612	[R]	633	[R]	683	[R]	714	[R]	741
24,253	[R]	24,823	[R]	26,064	[R]	27,453	[R]	28,485	[R]	29,195

ENERGY TIME LINE: 3000 B.C. TO A.D. 2009

3000 B.C.	Mesopotamians use petroleum for a range of purposes, including medicine, roads, shipbuilding, and architecture.
2800	Sales of olive oil for use as fuel in lamps and for cooking are recorded on clay tablets in Sumer.
1100	Written evidence of the use of coal for fuel appears in various localities.
200	China pioneers the use of natural gas as a fuel, developing a gas-fired evaporator used to extract salt from brine. Gas reaches the evaporators from shallow wells by means of simple percussion rigs and bamboo piping.
250–400 A.D.	Romans build a 16-wheel watermill in southern France, which produces more than 40 horsepower.
500–900	Persians invent the first windmills, using them to pump water and grind grain.
600	Middle Eastern chemists discover an incendiary weapon—comparable to modern napalm—derived from petroleum and quicklime.
874	Iceland is settled. Geothermal energy keeps the new inhabitants warm.
1400s	Coal becomes a viable fuel for common use in home heating because of the invention of firebricks, which make chimney construction inexpensive.

1626	French explorers document the burning of natural gas from seeps by Native Americans at Lake Erie.
1769	James Watt patents the steam engine.
1800s	Coal becomes the principal fuel used by steam-powered trains.
1800–1826	Humphrey Davy builds a battery-powered arc lamp. The first energy utility in the United States is founded. The relationship between electricity and magnetism is confirmed. The first electric motor is developed by Faraday. Ohms Law is published.
1816	Natural gas lights up the street lamps of Baltimore. Through the 19th century, natural gas—at that time still largely derived from coal, rather than extracted directly from the earth—is used extensively as a lighting fuel in North America and Europe.
1830–1839	Michael Faraday builds an induction dynamo based on the principles of electromagnetism, induction, generation, and transmission. The first industrial electric motors are built. The first fuel cell is designed.
1860	Auguste Mouchout demonstrates that solar radiation can be converted into mechanical power. Wood remains the primary fuel for cooking and heating and is also used for steam generation in industries and transportation.
1870–1880	Draft animals account for more than half of the total horsepower of all prime movers. The gas turbine is invented. The first combustion engine is designed to use alcohol, and gasoline is made. Edison Electric Light Co. (U.S.) and American Electric and Illuminating (Canada) are founded. The first commercial power station opens in San Francisco using brush generator and arc lights. Thomas Edison opens the first electricity-generating plant (in London) in January 1881. Edison's Pearl Street Station opens in New York as the first American plant to generate electricity. A month after beginning operations, it is feeding 1,300 light bulbs. Within a year, it is feeding 11,000 bulbs—each a hundred times brighter than a candle.

1878	William Adams constructs a reflector of flat-silvered mirrors, arranged in a semicircle, that concentrates solar radiation onto a stationary boiler.
1881–1887	The first hydroelectric station opens (Wisconsin). The transformer is invented. The steam turbine is invented. William Stanley develops the transformer and invents the alternating current electric system. Nicola Tesla invents the induction motor with a rotating magnetic field. This makes unit drives for machines and AC power transmission economically feasible. The electron is discovered.
1883	Charles Fritts builds the first solar cell.
1883–1884	John Ericsson (U.S.) invents and erects a solar engine using the parabolic trough construction.
1885	Robert Bunsen invents the "Bunsen burner," which produces a flame that can be safely used for cooking and heating with the mixing of the right proportion of natural gas and air.
1888	Charles F. Brush uses the first wind turbine to generate electricity in Cleveland, Ohio. Brush Electric Co. will ultimately be acquired by General Electric.
1890s	Electricity begins to replace natural gas for lighting purposes. Coal displaces much of the wood used in steam generation.
1900	Ethanol competes with gasoline to be the fuel for cars. Rudolph Diesel demonstrates his first engine. It runs on peanut oil.
1900–1910	The first geothermal electricity commercialization begins in Italy. The first electric vacuum cleaner is produced. The first electric washing machine is sold. Henry Ford's Model T is designed to use ethanol, gasoline, or any combination of the two fuels. The first pumped storage plant (Switzerland) opens. One of the most significant events of the 20th century is Albert Einstein's discovery of $E = mc^2$. This eventually leads to nuclear power, nuclear weapons, nuclear medicine, and astrophysics.
1906–1970	U.S. residential demand for natural gas grows 50 times bigger.

1910	Most rural homes are still heated with wood. In towns, coal is displacing wood in homes.
1920	The Ford Motor Company manufactures the Model T in large numbers.
1940s–1960s	Thousands of miles of new pipeline are constructed throughout the United States, leading to rapid growth in the natural gas market.
1942	The Manhattan Project is formed in the United States to secretly build the atomic bomb for use in World War II.
	The first controlled nuclear chain reaction is led by Enrico Fermi (U.S. immigrant from Italy) and other scientists at the University of Chicago.
1950	Electricity and natural gas displace wood heat in most homes and commercial buildings.
	Oil surpasses coal as the country's number one fuel source.
	Americans own 50 million cars.
1956	President Eisenhower signs the Federal-Aid Highway Act of 1956, which establishes the interstate highway system.
Mid-1950s	The Bridgers-Paxton Building, now listed in the National Historic Register as the world's first solar-heated office building, is designed.
1957	The first full-scale nuclear power plant (Shippingport, Pennsylvania) begins service.
1958	Airlines begin replacing propeller planes with jet planes.
1961	Coal has earned its place as the primary fuel for electricity generation in the United States.
1973	Several Arab OPEC nations embargo the sale of oil to the United States and Holland.
1986	The Perry power plant in Ohio becomes the 100th U.S. nuclear power plant in operation.
	The world's worst nuclear power accident happens at the Chernobyl plant in the former USSR (now Ukraine).
1987	Congress selects Yucca Mountain in Nevada for study as the first high-level nuclear waste repository site.
1990	More than 2,200 megawatts of wind energy capacity are installed in California—more than half of the world's capacity at the time.
	The Clean Air Act amendments require many changes to gasoline and diesel fuels to make them pollute less. The

use of these cleaner fuels is phased in during the 1990s. From 1995 on, "reformulated" gasoline is used in places with the worst pollution problems.

1993–forward For the first time, the United States imports more oil and refined products from other countries than it produces. More and more imports are needed because of growing petroleum demand and declining U.S. production.

1997 The Kyoto Protocol, an international agreement for industrialized nations to cut emissions by 5 percent by 2010, is adopted. The United States does not sign.

2005 Trucking accounts for 65 percent of energy used for transporting freight. Water transportation accounts for 18 percent, natural gas pipelines for 9 percent, and Class I railroads for 8 percent.

The Energy Independence and Security Act of 2007 sets a new corporate average fleet efficiency (CAFE) standard for cars and light trucks. The new standard will require car makers to meet a fleet-wide average of at least 35 miles per gallon by 2020, a 40 percent increase over the old standard.

The Energy Policy Act of 2005 is responsible for regulations that ensure gasoline sold in the United States contains a minimum volume of renewable fuel.

2007 U.S. wind power produces enough electricity on average to power the equivalent of more than 2.5 million homes. The installed capacity of wind-powered electricity-generating equipment is 13,885 megawatts as of September 30, more than four times the capacity in 2000.

Browns Ferry Nuclear Power Plant Unit 1 is the first U.S. nuclear reactor to come online in the 21st century.

2008 In the United States, crude oil price break $100 per barrel for the first time.

In the United States, gasoline prices break $4 per gallon for the first time.

2009 The American Recovery and Reinvestment Act includes billions of dollars for energy-efficiency and renewable-energy programs and research activities.

2010 China takes the lead as the world's largest manufacturer of wind turbines and solar panels.

PROFILES

ADAMS, WILLIAM

Designed solar panels, which tracked sunlight. The electricity was used to power engines for large-scale power plants.

BACON, FRANCIS T.

British scientist who built the first practical hydrogen–air fuel cell, which was used to power welding machines. NASA now uses Bacon's fuel cell for everyday needs and on spacecraft.

BECQUEREL, A. E.

French physicist who observed the photoelectric effect. He also measured intensity of light by using photochemical reactions.

BRUSH, CHARLES F.

Built the first windmill to generate power on a large scale in Cleveland, Ohio. His windmill had 144 blades and was 17 meters in diameter. His windmill design produced 12 kW of power, which he stored in batteries.

CLAUDE, GEORGE

Built the first system for harnessing energy from the oceans. This paved the way for Steven Salter, who works with ocean energy systems and is the inventor of the Salter duck. (See later entry for Salter.)

CONDOOR, SRIDHAR

St. Louis University mechanical engineer who developed the first hollow wind turbine. His development can supply up to 75 percent of the average home's energy needs. His turbine wraps around a chimney, tree, or utility pole and can catch breezes from any direction.

CONLOGUE, FRED

Director of design services for Hannaford Bros. supermarket chain who was instrumental in creating one of the first stores to meet LEED building standards.

CONRAD, WILLIAM

Conrad, an American, was the first person to pilot an airplane powered by hydrogen gas as the fuel.

DE SAUSSURE, HORACE BENEDICT

Swiss physicist and geologist who designed the first solar water heater, consisting of a wooden box with a black face and a glass top.

DRAKE, EDWIN

Drilled the first oil well in Titusville, Pennsylvania. The oil was refined through fractional distillation to make kerosene to be used in lamps and heaters.

EINSTEIN, ALBERT

Won the Nobel Prize in physics for his theories explaining the photoelectric effect. A. E. Becquerel observed the photoelectric effect while studying intensities of light.

ERICSSON, JOHN

Expanded on Mouchout's solar panel design using a parabolic trough instead of a dish, which became the standard for modern-day parabolic troughs.

ERREN, RUDOLF

Received patents for engines running on pure hydrogen. His Erren engines were used to run a fleet of industrial trucks and railroad cars.

FARADAY, MICHAEL

Discovered that a conductor moving through a magnetic field produces an electric current. In a hydroelectric plant, turbines provide rotational energy created by the kinetic energy of moving water. The rotational energy spins an armature in a coil of copper wire, generating electricity.

FERMI, ENRICO

Won the Nobel Prize in physics for his study of the decay of unstable isotope nuclei. He built the first "nuclear pile" under the football stands at the University of Chicago.

FRITTS, CHARLES

Constructed the first selenium solar cell. His design was inefficient, converting less than 1 percent of received light into usable electricity.

FULLER, BUCKMINSTER

Designer of a solar-powered geodesic dome house. He discovered Buckminster fullerene, a crystalline form of carbon similar to a geodesic dome.

FULLER, CALVIN

Bell scientist and the first to devise a semiconductor made of phosphorus and boron, increasing the efficiency of semiconductors to 15 percent.

GERDEMAN, FREDERICK

A Department of Energy biofuels expert who is experimenting with an open pond system for producing algae for biofuel.

GRANT, JOHN D.

Drilled a well in a place called The Geysers in California, creating the first geothermal power plant in the United States.

GROVE, WILLIAM-ROBERT

Devised an electric cell making use of hydrogen and oxygen to produce electricity as they combined to form water. His fuel cell is now known as a hydrogen fuel cell and was used in the spacecraft when NASA astronauts went to the moon.

HALLIDAY, DANIEL

A New Englander who designed a windmill with more than the usual four blades and with a vane orienting the blades to the wind. The blades were hinged so that they could fold up in extremely high winds to avoid damage.

KAZIMI, MUJID

Director of MIT's Center for Advanced Nuclear Systems. He says commercial reactors provide 20 percent of the United States' power but account for 70 percent of our emission-free energy.

MOUCHOUT, AUGUSTE

A French inventor who designed and patented a disk-shaped solar reflector that used solar rays to heat water to create steam to power a motor.

MUSK, ELON

South African–born owner of a new company, Tesla Motors. His goal is to develop a practical car that runs entirely on electricity. His company is named for Nikola Tesla, who studied ways to get free electricity from the atmosphere to power America.

NAUEN, ANDREAS

CEO of the Siemens wind power unit. The German company is a leading manufacturer of wind turbines, in the growing field of wind turbine energy.

PAUL, STEPHEN

Princeton thermonuclear physicist who was the first to use garbage as a substitute for gasoline. He calls it P (for Princeton) series fuel, which is a blend of 45 percent ethanol, 35 percent natural gas, and 20 percent methyltetrahydrofuran (MeTHF).

SALTER, STEVEN

Mechanical engineer who works with ocean energy systems. Inventor of the Salter duck, a series of flaps, which pivot around a shaft, driving a hydraulic fluid to produce electricity.

SELSAM, DOUGLAS

Inventor of a wind turbine called the Sky Serpent. His wind turbine is so compact that it can be carried by hand and adapted for many commercial uses.

THACKERAY, MICHAEL

A battery expert working at Argonne National Laboratory. His mission is to develop a next-generation electric battery that will meet today's strategic and industrial requirements.

OPPORTUNITIES IN RENEWABLE AND NONRENEWABLE ENERGY CAREERS

AMERICAN SOLAR ENERGY SOCIETY—WWW.ASES.ORG

This site presents green-collar jobs forecast in the United States to the year 2030. It explores job opportunities in wind, solar, thermal, photovoltaics, fuel cells, and biofuels.

CLEAN EDGE JOBS—WWW.JOBS.CLEANEDGE.COM

This is a source of job listings for clean technology job seekers, employers, and recruiters.

CLEAN LOOP—WWW.CLEANLOOP.COM

Clean Loop lists job opportunities in emerging for-profit companies that are exploring new technologies to create alternate fuel sources and software applications to "revolutionize" the energy industry.

CLEAN TECHNOLOGY JOBS—WWW.TECHNICALGREEN.NET

This is a green job locater and network for career opportunities in renewable energy, sustainable agriculture, and green building technology.

EERE, U.S. DEPARTMENT OF ENERGY— WWW1.EERE.ENERGY.GOV

Provides information on clean energy jobs in the public, private, and nonprofit sectors, ranging from entry-level opportunities to professional positions.

ENERGY CAREERS—WWW.ISEEK.ORG

A site that explores the question, "What energy career is right for you?" It focuses on examining career opportunities in engineering, installation and repair, production, and construction.

ENVIRONMENTAL GREEN CAREERS CENTER— WWW.GREENCAREERS.COM

This site offers a comprehensive listing of environmental and natural resources job opportunities, with a focus on career news, inside tips and advice for job seekers, and career research reports.

GET INTO ENERGY—WWW.GETINTOENERGY.COM

The focus of this site is to develop an awareness among students, parents, and educators regarding career paths in the energy industry.

GREEN BIZ—WWW.GREENBIZ.COM

This site maintains a list of job postings and internships for green jobs in solar and renewable energy, clean tech, green building, and sustainable businesses.

GREEN CAREERS GUIDE— WWW.GREENCAREERSGUIDE.COM

This database displays articles on green jobs and presents career guidance on jobs, training, and green entrepreneurship.

GREEN CAREERS JOURNAL— WWW.ENVIRONMENTALCAREER.COM

This is a publication that contains current green jobs listings and information as well as articles on environmental careers and a growing green economy.

GREEN CORPS—WWW.GREENCORPS.ORG

Green Corps offers hands-on experiences and training for university graduate students to help them find careers with organizations committed to resolving global environmental issues.

GREEN DREAM JOBS— WWW.SUSTAINABLEBUSINESS.COM

A sustainable business job service that posts renewable energy jobs in solar, wind, geothermal, and wave energy and green building technology, as well as opportunities in government green-job areas.

GREEN ENERGY JOBS— WWW.GREENENERGYJOBS.COM

Provides a career guide to those wanting an overview of opportunities in renewable resources: green building, planning, marine energy, wave energy, hydro energy, bioenergy, solar technology, and micro-renewable energy.

GREEN JOBS NETWORK— WWW.GREENJOBS.NET

The goal of the network is to connect people seeking jobs that focus on environmental and social responsibilities to available related opportunities and services.

TREE HUGGER JOB BOARD— WWW.JOBS.TREEHUGGER.COM

The job board lists recent green and non-green jobs in a variety of occupational categories related to environmental sustainability.

U.S. DEPARTMENT OF ENERGY, CAREER OPPORTUNITIES— WWW.DOE.GOV

Features information about job vacancies in the U.S. Department of Energy and its DOE laboratories.

U.S. DEPARTMENT OF LABOR, CAREER VOYAGES— WWW.CAREERVOYAGES.GOV

This is a site that explores job training opportunities available in various renewable energy industries.

U.S. GREEN BUILDING COUNCIL—WWW.USGBC.ORG

A career center established to connect applicants to employment opportunities in green job technology.

USA GREEN ENERGY JOBS— WWW.USAGREENENERGYJOBS.COM

This site presents a sample of green energy jobs by type in states and cities in the United States.

VOCATIONAL INFORMATION CENTER— WWW.KAHKE.COM

An international site that explores careers in energy with links to skill requirements, salary, and training and job opportunities.

ENERGY PRODUCT DEVELOPERS AND MANUFACTURERS

Besides the following product developers and manufacturers, you can also go to an online buyer's guide and business directory for renewable energy businesses and organizations worldwide: www.energy.sourceguides.com

ABENGOA SOLAR, DENVER, CO

Develops and constructs solar power tower systems and photovoltaic cells for use in the production of electricity. www.abengoasolar.com

ABUNDANT RENEWABLE ENERGY, NEWBURG, OR

Manufactures wind energy generators and towers designed for harsh climates and low wind-speed areas. www.abundantre.com

ALTA ROCK ENERGY INC., SEATTLE, WA

Develops and commercializes geothermal deep drilling technology. www.altarockenergy.com

AUTOMOTIVE INDUSTRY RESEARCH, INTERNATIONAL

Automobile manufacturers are exploring engineering strategies to produce clean and efficient vehicles using biofuels, tire and motor oil technology, hydrogen fuel cells, lithium-ion battery technology, fuel-efficiency technology, and light plastic materials. www.cargroup.org

BP PETROLEUM, WARRENVILLE, IL

Developed a carbon capture and storage technology that extracts carbon emissions from fossil fuels and processes them into hydrogen to generate electricity and capture and store carbon elements permanently underground. www.BP.com/EnergyLab

BRIGHT SOURCE ENERGY, OAKLAND, CA

Builds, owns, and operates large-scale solar energy projects. www.brightsourceenergy.com

CARRIER CORPORATION, FARMINGTON, CT

Manufactures geothermal heat pumps for use in residential heating and cooling systems. www.residentialcarrier.com

CETC SOLAR GROUP, CHANGSHA, CHINA

Manufacturer and supplier of all solar products, including solar cells and panels and photovoltaic systems. www.cetc-solar.com

CHEVRON ENERGY SOLUTIONS CO., SAN FRANCISCO, CA

Applies proven energy-efficiency and renewable-power technologies such as infrastructure systems, energy controls, solar power, biomass, and fuel cells to meet the facility needs of individual and institutional customers. www.chevron.com/globalissues

E.I. DUPONT DE NEMOURS, BREVARD, NC

Manufactures alternate fuel boilers, which convert on-site industrial waste materials and nonrecyclable by-products into usable steam energy. www2.dupont.com

EXXON MOBIL, HOUSTON, TX

Designs and uses equipment for extracting oil and gas reserves while reducing the environmental impact of energy development. www.exxonmobil.com

FRAUNHOFER INSTITUTE FOR SOLAR ENERGY SYSTEMS, FEIBURG, GERMANY

Research and production of solar electric power systems and photovoltaic modules. www.fraunhofer.de

GENERAL ELECTRIC, ATLANTA, GA

Manufactures products for the energy industry incorporating the use of fossil fuels, nuclear, solar, and wind applications. www.gepower.com

IBM, SAN JOSE, CA

Using nano-membrane technology, it is developing lightweight, high-energy lithium air batteries. www.almadenibm.com

NANOSOLAR, SAN JOSE, CA

Developed the Nanosolar Utility Panel, the first designed and manufactured solar electricity panel for inclusion in utility-scale solar powered systems. www.nanosolar.com

NEVADA SOLAR ONE, BOULDER CITY, NV

Constructed and maintains a solar energy plant that concentrates and converts desert sunlight into thermal energy for electric power generation. www.acciona-na.com

OERLIKON SOLAR, SWITZERLAND

Mass-produces thin-film silicon solar modules. www.oerlikon.com

OXFORD YASA MOTORS, GREAT BRITAIN

Manufactures lightweight, energy-efficient electric motors for the automobile industry. www.ox.ac.uk

PV CRYSTALOX SOLAR, ERFURT, GERMANY

Manufactures photovoltaic cell materials, solar-grade silicon, silicon wafers, and ingots. www.pvcrystalox.com

SANDIA NATIONAL LABORATORIES, LIVERMORE, CA

Researches and develops commercially viable energy technologies based on wind, solar, and geothermal resources. www.public.ca.sandia.gov

SIEMENS CORPORATION, NEW YORK, NY

Manufactures wind turbines for onshore, coastal, and offshore sites. www.energysiemens.com

SOLIX BIOFUELS, COYOTE GULCH, CO

Planned and built a demonstration facility that is anticipated to produce 3,000 gallons of algal biofuels per acre per year. www.solixbiofuels.com

SUNCOR (SUNOCO) ENERGY INC., ALBERTA, CANADA

Maintains an ethanol facility with a capacity to produce 200 million liters per year. The refined ethanol is blended into gasoline products. www.suncor.com

USDA SOUTHERN RESEARCH STATION, ASHEVILLE, NC

Partners with private industries to use basic and applied science to develop wood energy products from southern forests. www.srs.fs.usda.gov

XTREME POWER AND CLAIRVOYANT ENERGY, DEARBORN, MI

These companies have converted an idle Ford Motor Company assembly plant into one of the nation's largest renewable-energy manufacturing parks. They produce solar power and energy storage systems. www.xtremepowerinc.com

NATIONAL SCIENCE EDUCATION STANDARDS, CONTENT STANDARDS

Unifying Concepts and Processes, K–12
Systems, order, and organization
Evidence, models, and explanation
Constancy, change, and measurement
Evolution and equilibrium
Form and function

Science as Inquiry, Content Standard A, Grades 9–12
Abilities necessary to do scientific inquiry
Understandings about scientific inquiry

Physical Science, Content Standard B, Grades 9–12
Structure of atoms
Structure and properties of matter
Chemical reactions
Motions and forces
Conservation of energy and increase in disorder
Interactions of energy and matter

Life Science, Content Standard C, Grades 9–12
The cell
Molecular basis of heredity
Biological evolution
Interdependence of organisms
Matter, energy, and organization in living systems
Behavior of organisms

Earth and Space Science, Content Standard D, Grades 9–12
Energy in the earth system
Geochemical cycles
Origin and evolution of the earth system
Origin and evolution of the universe

Science and Technology, Content Standard E, Grades 9–12
Abilities of technological design
Understandings about science and technology

Science in Personal and Social Perspectives, Content Standard F, Grades 9–12
Personal and community health
Population growth
Natural resources
Environmental quality
Natural and human-induced hazards
Science and technology in local, national, and global challenges

History and Nature of Science, Content Standard G, Grades 9–12
Science as a human endeavor
Nature of scientific knowledge
History of science

INDEX

Boldface page numbers refer to volume numbers. A key appears on all verso pages. An italicized *t* following a page number indicates a table. An italicized *f* following a page number indicates a figure.

A.A. Kingston Middle School, **2**:17, **5**:34, **5**:34*f*
Abate, Dee, **5**:39
Abengoa Solar, **1**:213, **2**:40, **2**:55, **2**:183, **3**:181, **4**:183, **5**:185
ABI. *See* Allied Business Intelligence
Abu Dhabi, **5**:92–93, **5**:93*f,* **5**:129
Abundant Renewable Energy, **1**:213, **2**:183, **3**:181, **4**:183, **5**:185
ACC. *See* American Coal Council
Acciona's Solar One, **2**:38*f*
Acid mine drainage (AMD), **1**:109–11, **1**:110*f*
Active solar heating systems, **2**:70–74
Active solar water heaters, **2**:76*f*
Active yawing, **3**:12
Adams, William, **1**:198, **1**:203, **2**:5, **2**:168, **2**:173, **3**:166, **3**:171, **4**:168, **4**:173, **5**:170, **5**:175

Adlai E. Stevenson High School, **5**:38*f*
Advanced DC 4001 30 HP electric motor, **2**:96
Aeroturbine, **3**:14
Afghanistan, **3**:94
Africa: OTEC station off of, **3**:122; solar energy used in, **2**:26
Ahuachapán geothermal field, **4**:43
Airborne wind turbines, **3**:62
Aircraft propulsion, **2**:105
Air quality standards, **4**:25–26, **5**:29–30
Air-to-water heat pumps, **4**:74
Alamos National Laboratory, **5**:115
Alaska: geothermal resources of, **4**:18–19; as oil-producing state, **1**:55*f;* pipeline, **1**:54

1: Oil, Natural Gas, Coal, and Nuclear
2: Solar Energy and Hydrogen Fuel Cells
3: Wind Energy, Oceanic Energy, and Hydropower
4: Geothermal and Biomass Energy
5: Energy Efficiency, Conservation, and Sustainability

Alaska North Slope, **1**:83
Albuquerque, New Mexico, **5**:17
Aleman, Angel, **1**:50
Aleutian Islands, **4**:18
Algae, **2**:93*f*; as biofuel, **4**:115; blue-green, **2**:94; high-oil, **2**:93; hydrogen produced by, **2**:92–94
All American Homes, **5**:23
Alliance to Save Energy, **1**:27, **5**:42
Allied Business Intelligence (ABI), **2**:117, **5**:118
Alonzo, Stephanie, **1**:50
Altamont Pass Wind Farm, **3**:32, **3**:33*f*
Alta Rock Energy, **1**:213, **2**:183, **3**:181, **4**:183, **5**:185
Alternate Fuels and Advanced Vehicles Data Center, **5**:127
Alternative energy, **5**:59–62
Alternative Energy Primer, **4**:105
Alternative Fuels and Advanced Data Center, **2**:103, **4**:105
Aluminum industry, **1**:21
AMD. *See* Acid mine drainage
American Coal Council (ACC), **1**:35, **1**:129
American Electric and Illuminating, **1**:198, **2**:168, **3**:166, **4**:168, **5**:170
American Federation of Teachers, **5**:37
American Gas Association, **1**:35, **1**:100, **1**:165, **2**:135, **3**:133, **4**:135, **5**:137

American Geologic Institute, **1**:68
American Geophysical Union, **1**:68
American Hydrogen Association, **2**:127
American Institute of Architects, **5**:91
American Nuclear Society, **1**:35, **1**:157, **1**:165, **2**:135, **3**:133, **4**:135, **5**:137
American Petroleum Institute, **1**:35, **1**:66
American Recovery and Reinvestment Act, **1**:33, **1**:201, **2**:171, **3**:169, **4**:89, **4**:171, **5**:120, **5**:173
American Solar Energy Society (ASES), **1**:35, **1**:165, **1**:209, **2**:33, **2**:80, **2**:135, **2**:179, **3**:133, **3**:177, **4**:135, **4**:179, **5**:137, **5**:181
American Wind Association, **1**:35
American Wind Energy Association (AWEA), **1**:165, **2**:135, **3**:2, **3**:26, **3**:133, **4**:135, **5**:137
Anaerobic digestion, **4**:97
Animal husbandry, **4**:100
Anode, **2**:88
Antifreeze, **4**:60, **4**:66*f*
ANWR. *See* Arctic National Wildlife Refuge
Appliances, **5**:50*f*; efficiency of, **5**:52–53; fuel use and, **1**:178*t*–179*t*, **2**:148*t*–149*t*, **3**:146*t*–147*t*, **4**:148*t*–149*t*, **5**:150*t*–151*t*
AquaBuoy, **3**:116*f*, **3**:117*f*
Aquaculture, **4**:50
Aramaki, Teiichi, **1**:32*f*
Arch dam, **3**:80
Archimedes, **2**:35
Architecture: ecological, **5**:24; green, **5**:88–89; for green roof, **5**:81–82
Arctic ice mass, **1**:24*f*
Arctic National Wildlife Refuge (ANWR), **1**:61
Arizona: geothermal energy in, **4**:18; geothermal heat pumps tested in,

4:65–66; solar power plant in, **2**:39; Tucson, **5**:17
Arkansas, **4**:5, **4**:118, **5**:91, **5**:91*f*
Arquin, Michael, **3**:47–51, **3**:48*f*
Arsdell, Brent Van, **2**:40*f*
Arsene d'Arsonval, Jacques, **3**:118
ASES. *See* American Solar Energy Society
Association for the Advancement of Sustainability in Higher Education, **5**:41
Association of American State Geologists, **1**:67
Atoms, **2**:10
Auburn University, **4**:83
Austin, Texas, **5**:91
Australia, **1**:85; coal exports of, **1**:116; geothermal power plants in, **4**:42; hot dry rock resources in, **4**:47–48; photovoltaic technology used by, **2**:25
Austria: biomass energy in, **4**:89; hydroelectric power plants in, **3**:89
AutoDesk, **5**:18
Automotive Industry Research, **1**:213–14, **2**:183–84, **3**:181–82, **4**:183–84, **5**:185–86
AWEA. *See* American Wind Energy Association
Aydil, Eray, **5**:114*f*

B20 fuel, **4**:119*f*, **4**:120
Babcock Ranch, **2**:22
Backhus, DeWayne, **3**:20
Bacon, Francis T., **1**:203, **2**:90, **2**:173, **3**:171, **4**:173, **5**:175
Bacteria, **4**:103
Baez, Ana, **1**:50
Bahrain World Trade Center, **3**:55–56, **3**:55*f*
Ballard Power System, **2**:90, **2**:102
Ball State University (BSU), **4**:68
Bantam, Doug, **4**:1
Bargeloads, of coal, **1**:113*f*

Baring-Gould, Ian, **3**:44
Barrage technologies, **3**:107
Basics of Energy Efficient Living (Wibberding), **5**:61
Bates, John, **5**:123
Bats, **3**:60
Batteries: fuel cells using, **2**:97; future of, **5**:122–23; NiMH, **5**:122–23; thin-film lithium-ion, **5**:123, **5**:123*f*
Battersdy, Leah, **5**:64*f*
Battery storage: for home and business, **2**:15*f*; PV to, **2**:14
Bay Localize, **1**:28
Bay of Fundy, **3**:110
Beaufort, Francis, **3**:5, **3**:6*t*
Beaufort scale, **3**:5, **3**:6*t*
Beaver County power plant, **4**:20
Becquerel, Edmond, **1**:203, **2**:2, **2**:173, **3**:171, **4**:173, **5**:175
Belize, **4**:34
Bell Laboratories, **2**:5
Belote, Dave, **2**:3*f*
Benz, Daimler, **2**:90
Benzene, **1**:53
Bergey Windpower, **3**:72
Berkeley Biodiesel Collective, **4**:126
Biliran, **4**:33
Billings, Montana, **5**:91
Binary power plants, **4**:9–10, **4**:9*f*, **4**:22, **4**:40*f*
Binder, Michael, **2**:98
Biodiesel, **1**:48, **1**:49, **1**:51, **1**:53–54, **4**:127–28; advantages of, **4**:121; of America, **4**:126; Arkansas school buses using, **4**:118; B20 fuel, **4**:119*f*, **4**:120; buses using, **4**:109*f*; California using, **4**:121; composition of, **4**:112; defining, **4**:109; disadvantages of, **4**:121–25; discarded restaurant oil used as, **4**:116*f*; grassroots effort in, **4**:112; high-oil algae for, **2**:93; home heating with, **4**:120; Idaho projects of, **4**:118; petroleum diesel

1: Oil, Natural Gas, Coal, and Nuclear
2: Solar Energy and Hydrogen Fuel Cells
3: Wind Energy, Oceanic Energy, and Hydropower
4: Geothermal and Biomass Energy
5: Energy Efficiency, Conservation, and Sustainability

emissions compared to, **4**:116; production specifications of, **4**:114–16; race car using, **4**:107–8, **4**:108*f*; school buses using, **4**:117–20, **4**:119, **4**:119*f*; students building cars using, **4**:121; in US, **5**:122; vegetable oil as, **4**:116*f*; vehicles powered by, **4**:110–11, **4**:117–20, **5**:75
Biodiesel Solutions, **4**:107
Biodigester, **4**:98, **4**:99*f*, **4**:101
Bioethanol, **4**:91–92
Biofuel, **1**:xvii, **1**:14, **2**:xvii, **3**:xvii, **4**:xvii, **4**:89–90, **5**:xvii; algae used as, **4**:115; bacteria producing, **4**:103; biotechnology used in, **5**:123; buses, **5**:7, **5**:8*f*; defining, **4**:90–91; dry-milling for, **4**:94*f*; in sustainable development, **5**:120–22; US consumption of, **1**:61
Biogas, **4**:96–97, **4**:128; China's use of, **4**:102–3; CO$_2$ from, **4**:96; cow manure producing, **4**:98; digester, **4**:100–101; production facilities for, **4**:97*f*; savings from, **4**:101–2; technology of, **4**:102–3
Biogas plant, components of, **4**:98
Biogen Idec, Inc., **4**:64
Biojet, **4**:107
Biomass, **1**:13–14, **2**:64–65; Asian countries using, **4**:85–88; Austria's energy source of, **4**:89; benefits of, **4**:103–4; defining, **4**:81–82; Denmark's energy source of, **4**:89; economic benefits of, **4**:104; electricity capacity of, **4**:104; Finland using, **4**:88; gasification plant, **4**:79, **4**:80*f*; heating system using, **5**:25; hybrid poplars as, **4**:83–85, **4**:84*f*; Indonesia's energy source of, **4**:86; landfill gas from, **1**:97; Philippines' energy from, **4**:87; reading materials on, **1**:162, **2**:132, **3**:130, **4**:132, **5**:134; renewed interest in, **4**:85; Sweden's energy source of, **4**:88; switchgrass as, **4**:82–83, **4**:82*f*; types of, **4**:81*f*; US government interest in, **4**:80; US percentage use of, **4**:85; Vietnam's energy source of, **4**:87; wood-burning boilers and, **4**:105
Biomass Research Center, **4**:105
Biomass Solar Greenhouse Project, **2**:64
Bioreactors, **2**:93*f*
Biorefinery plants, **4**:90*f*
Biotechnology, **5**:123
Birdsville geothermal power plant, **4**:42
Bitumen, **1**:58, **1**:59
Blackfeet Indian Reservation, **5**:25–26
Blade design, **3**:9
Blenders, **1**:53–54
Blohm, Margaret, **5**:117*f*
Bloom Energy Corporation, **5**:121–22, **5**:121*f*, **5**:128
Bloom Energy Server, **5**:121–22
Blower door test, **5**:56*f*
Blue-green algae, **2**:94
Bluenergy Solarwind Turbine, Inc., **3**:14
Blue Sun Company, **4**:115
BMW Mini E, **5**:72
Boeing Research & Technology, **2**:105, **2**:106*f*
Boiling water reactors, **1**:142–43, **1**:143*f*

Boise State University, **3:**46
Bolluyt, Jan, **3:**19–23
Bonneville Dam, **3:**97*f*
Borrego Solar Systems, **2:**33
Boston, Massachusetts, **5:**91
Boulder, Colorado, **5:**91
BP. *See* British Petroleum
Brazil, **2:**113, **3:**86–87
Breakthrough Technologies Institute, **2:**103
Breeder reactors, **1:**143–44; fast, **1:**149–50; liquid metal fast, **1:**150*f*
Bright Source Energy, **1:**214, **2:**184, **3:**182, **4:**184, **5:**186
British Petroleum (BP), **1:**67, **1:**214, **2:**184, **3:**182, **4:**184, **5:**186
British thermal unit (Btu), **1:**18, **1:**78
Browning High School, **5:**25–26
Browns Ferry Nuclear Power Plant, **1:**201, **2:**171, **3:**169, **4:**171, **5:**173
Brush, Charles F., **1:**199, **1:**203, **2:**169, **2:**173, **3:**167, **3:**171, **4:**169, **4:**173, **5:**171, **5:**175
BSU. *See* Ball State University
Btu. *See* British thermal unit
Buckley Air Force Base, **2:**14
Buffalo Ridge Wind Farm, **3:**34
Bunsen, Robert, **1:**199, **2:**169, **3:**167, **4:**169, **5:**171
Burdin, Claude, **3:**7
Buses: biodiesel, **4:**117–20, **4:**119*f*; biofueled, **5:**7, **5:**8*f*; CNG powered, **1:**86*f*; diesel-fueled, **1:**48*f*; hydrogen fuel cells, **2:**111–13, **2:**112*f*; soybean-powered, **4:**109*f*
Businesses: battery storage for, **2:**15*f*; carbon footprint of, **5:**18–19; wind farms, **3:**39–40
Butane, **1:**74, **1:**76
Buttress dam, **3:**80

C. reinhartii, **2:**92
CAA. *See* Clean Air Act
CAC. *See* Clean Air Council

C_aCO_3. *See* Calcium carbonate
Cactus Shadows High School, **4:**66
CAFE. *See* Corporate aver fleet efficiency
Caithness/COC, **4:**20
Calcium carbonate (C_aCO_3), **1:**111
CalEnergy Navy I, **4:**11
California: biodiesel used in, **4:**121; Energy, **4:**20; geothermal energy in, **4:**14–15; go-green projects in, **5:**33–34; Golden Gate Bridge and, **3:**111, **3:**112*f*; solar energy used in, **2:**17; wind energy in, **1:**200, **2:**170, **3:**168, **4:**170, **5:**172
California Fuel Cell Partnership, **2:**115–16
California Wind Energy Association, **3:**32
California Youth Energy Services (CYES), **1:**30
Calorie, **1:**18
Calpine, **4:**6, **4:**20
Canada: airborne wind turbines in, **3:**62; energy consumption per capita of, **1:**22; geothermal activity in, **4:**38–39; hydroelectric generation of, **3:**85–86; hydrogen fuel cell buses in, **2:**111
Canadian Hydrogen Highway, **2:**111
Canola, **4:**115
Cantor, Phillip, **4:**122–25
Caprocks, **1:**74
Carbon, **1:**94, **5:**7
Carbon capture and sequestration, **1:**122–23, **1:**124*f*
Carbon cycle, **4:**91*f*
Carbon dioxide (CO_2), **1:**75, **4:**26; from biogas, **4:**96; from coal, **1:**119–20; coal-fired power station capture and storage of, **1:**124*f*; countries with highest emissions of, **5:**5; deforestation increasing, **1:**120; electricity and reduction of, **5:**48–50; emissions, **1:**95*f*;

> 1: Oil, Natural Gas, Coal, and Nuclear
> 2: Solar Energy and Hydrogen Fuel Cells
> 3: Wind Energy, Oceanic Energy, and Hydropower
> 4: Geothermal and Biomass Energy
> 5: Energy Efficiency, Conservation, and Sustainability

1:119–20, **1**:120*f*; emissions increasing of, **5**:4–5; emissions of fuel, **1**:95*f*; gas injection using, **1**:45; geothermal energy credits for, **4**:51; geothermal heat pump reducing, **4**:68, **4**:69; global emissions of, **1**:192–95, **2**:162–65, **3**:160–63, **4**:162–65, **5**:5, **5**:105, **5**:164–67; as greenhouse gas, **1**:23; long-term storage of, **1**:122–23; power plant producing, **1**:73*f*; trees capturing, **5**:1; US emissions of, **1**:120*f*; waste creating, **5**:51

Carbon footprint, **5**:5–6, **5**:104; of businesses, **5**:18–19; of cities, **5**:15–16; cities ranked for, **5**:17; cities reducing, **5**:17; emissions tracking of, **5**:18; estimating your, **5**:20; of homes, **5**:8–9; Idaho reducing, **5**:11–12; individuals, **5**:6*f*; legislative efforts reducing, **5**:21; Minnesota reducing, **5**:9–10; musicians reducing, **5**:6–8, **5**:8*f*; New Hampshire reducing, **5**:10–11; reducing, **5**:19; of schools and colleges, **5**:9–12; schools reduction of, **5**:12, **5**:14–15; Texas reducing, **5**:11; two parts of, **5**:6; Virginia reducing, **5**:11; Washington reducing, **5**:11

Carbon monoxide, **4**:92

Career resources, **1**:209–12, **2**:179–82, **3**:177–80, **4**:179–82, **5**:181–84

Carlisle, Anthony, **2**:90
Carlson, Jason, **5**:38*f*
Car maintenance, **5**:75–76
Carmichael, Don, **5**:37–40, **5**:38*f*
Carrier Corp., **1**:214, **2**:184, **3**:182, **4**:184, **5**:186
Catalytic filters, **5**:124
Cathode, **2**:88
Cattle, **1**:76
Cavendish, Henry, **2**:90
CDM. *See* Clean Development Mechanism
Cendejas, Emily, **1**:50
Central America, **1**:84
Certification plaque, **5**:81*f*
Cervantes, Janneth, **1**:50
CETC Solar Group, **1**:214, **2**:184, **3**:182, **4**:184, **5**:186
CFCs. *See* Chlorofluorocarbons
CFL. *See* Compact fluorescent light bulb
Chaplin, Daryl, **2**:5
Charcoal, **4**:102
Charest, Chris, **2**:95*f*
Charging station nozzle, **5**:73*f*
Chaudes-Aigues, France, **4**:32
Chemical energy, **1**:5
Chemical injection, **1**:46
Chen, David, **2**:52–54, **2**:52*f*
Chena Hot Springs, **4**:19
Chernobyl plant, **1**:200, **2**:170, **3**:168, **4**:170, **5**:172
Chevron, **1**:67, **4**:41
Chevron Energy Solutions Co., **1**:214, **2**:184, **3**:182, **4**:184, **5**:186
Chevy Volt, **5**:71
Chicago Biofuels, **4**:124
Chief Joseph Dam, **3**:75
China: biogas use of, **4**:102–3; coal-fired power station dependence reduced by, **4**:39; coal production of, **1**:114, **1**:116; Dongtan, **5**:94; energy use of, **5**:107; geothermal resources in, **4**:39; natural gas

extracted by, **1:**71; natural gas use by, **1:**197, **2:**167, **3:**165, **4:**167, **5:**169; Nuclear reactors in, **1:**11*f;* parabolic cookers used in, **2:**72; renewable energy promoted by, **3:**61*f;* rooftop solar heaters in, **2:**79*f;* solar cell manufacturing of, **1:**xiv, **1:**201, **2:**xiv, **2:**24, **2:**171, **3:**xiv, **3:**169, **4:**xiv, **4:**171, **5:**xiv, **5:**173; Three Gorges dam project of, **3:**83–85, **3:**84*f;* tidal power in, **3:**110; using microhydroelectric power plants, **3:**94; wind turbines in, **1:**201, **2:**171, **3:**60, **3:**61*f,* **3:**169, **4:**171, **5:**173

China Dome digester, **4:**102, **4:**103*f*

Chinese Guorui Biogas Company, **4:**102

Chlorofluorocarbons (CFCs), **1:**74

Chrysler ecoVoyager, **5:**70

Chu, Steven, **1:**125, **1:**131, **2:**85, **3:**73, **4:**80, **4:**89, **5:**108, **5:**111, **5:**120, **5:**129

Churchill, Winston, **3:**86

Churchill Falls, **3:**86

Cities: carbon footprint ranking of, **5:**17; carbon footprint reduction of, **5:**17; carbon footprints of, **5:**15–16; Solar America, **2:**21; using solar energy, **2:**17–21

Claude, Georges, **1:**204, **2:**174, **3:**118, **3:**172, **4:**174, **5:**176

CLC. *See* College of Lake County

Clean Air Act (CAA), **1:**47, **1:**114, **1:**122, **1:**201, **2:**171, **3:**169, **4:**171, **5:**173

Clean Air Act Amendments, **4:**115–16

Clean Air Council (CAC), **1:**165, **2:**135, **3:**133, **4:**135, **5:**137

Clean coal technology, **1:**121–28

Clean Coal Technology Program, **1:**122

Clean Development Mechanism (CDM), **4:**51

Clean Edge jobs, **1:**209, **2:**179, **3:**177, **4:**179, **5:**181

Clean Fuels Development Coalition, **4:**126

Clean Urban Transport for Europe (CUTE), **2:**112

Climate change, **5:**14–15, **5:**16*f;* environmental concerns for, **1:**22–24; reducing, **1:**25–33; United Nations Convention on, **1:**32*f*

Climate Protection Summit, **5:**16*f*

Clinton Climate Initiative, **5:**18

Closed-cycle systems, **3:**119–20, **4:**26

Closed-loop ground-coupled heat pump (GCHP), **4:**68

Closed-loop systems, **4:**60–61, **4:**61*f*

CNG. *See* Compressed natural gas

CO_2. *See* Carbon dioxide

Coal, **1:**xii, **1:**10, **2:**xii, **3:**xii, **4:**xii, **5:**xii; ash slurry, **1:**112; bargeload of, **1:**113*f;* carbon ratio of, **1:**94; China's production of, **1:**114, **1:**116; CO_2 from, **1:**119–20; consumption of, **1:**118; countries production of, **1:**115–16; deposits, **1:**106; early uses of, **1:**197, **2:**167, **3:**165, **4:**167, **5:**169; environmental issues of, **1:**109–11, **1:**128–29; exporters of, **1:**116–17, **1:**117*t;* formation of, **1:**106, **1:**107*f;* as fossil fuel, **1:**106; future of, **1:**128–29; gasification, **1:**125–28, **1:**126*f;* Germany and, **1:**115–16, **5:**106; history of, **1:**104; imports of, **1:**117; India's production of, **1:**116; Industrial Revolution and, **1:**10, **1:**115; mining of, **1:**106–12; Poland's production of, **1:**116; production, **1:**115–16; products made from, **1:**105*f;* recoverable reserves of, **1:**190*t*–191*t,* **2:**160*t*–161*t,*

1: Oil, Natural Gas, Coal, and Nuclear
2: Solar Energy and Hydrogen Fuel Cells
3: Wind Energy, Oceanic Energy, and Hydropower
4: Geothermal and Biomass Energy
5: Energy Efficiency, Conservation, and Sustainability

3:158*t*–159*t*, 4:160*t*–161*t*, 5:162*t*–163*t*; states producing, 1:114–15; sulfur dioxide from, 1:119; surface mining of, 1:109; transportation of, 1:112–13; uses of, 1:104–5; US industry of, 1:113–15, 1:115*f*, 1:117; Utah state rock as, 1:109; videos on, 1:130; world consumption of, 1:103

Coal-fired power stations: carbon dioxide capture and storage at, 1:124*f*; China reducing dependence on, 4:39; electricity produced by, 1:113–14, 1:114*f*; emissions from, 1:122, 5:124; environmental issues of, 1:119–20; nanotechnology based catalytic filters for, 5:124; zero emissions from, 1:122

Coal-generating plants, 1:121

Cob Connection, 4:124

College of Lake County (CLC), 5:39

Colleges: carbon footprints of, 5:9–12; Contra Costa Community, 2:4; green architecture in, 5:88–89; Iowa Central Community, 4:71; Middlebury, 4:79–80, 4:80*f*; solar energy installation at, 2:4; Texas State Technical, 3:47

Collins, Patrick, 5:125–26, 5:126*f*

Colorado: Boulder, 5:91; geothermal energy in, 4:17, 4:51*f*; wind farms in, 3:34

Columbia, 1:101

Compact fluorescent light bulb (CFL), 5:7, 5:53*f*

Components: of biogas plant, 4:98; of green buildings, 5:79; of green roofs, 5:84–86, 5:85*f*; of hydroelectric power plants, 3:81–83; of microhydroelectric power plants, 3:90–93, 3:92*f*; of wind turbines, 3:10

Compressed air storage, 2:28; for wind energy, 3:69–70; for wind farms, 3:31–32

Compressed natural gas (CNG): buses powered by, 1:86*f*; Egypt using, 1:93*f*

Computers, 5:54

Concentrating solar power (CSP), 1:xiv, 1:12, 2:xiv, 2:1, 2:35, 2:38*f*, 3:xiv, 4:xiv, 5:xiv; environmental impacts of, 2:50–51; future of, 2:50–51; land use of, 2:50; sustainable development with, 5:113; types of, 2:36–47

Concrete dome homes, 5:46, 5:47*f*

Condoor, Sridhar, 1:204, 2:174, 3:172, 4:174, 5:176

Conduction, 2:68

Congressional Research Service (CRS), 3:24, 3:30

Conlogue, Fred, 1:204, 2:174, 3:172, 4:174, 5:176

ConocoPhillips Company, 1:64–65, 1:67, 3:67

Conrad, William, 1:204, 2:174, 3:172, 4:174, 5:176

Conserv Fuels, 4:109

Construction materials, 5:80–81

Consumer Energy Center, 2:55

Consumers: of natural gas, 1:82–85; wind energy cost to, 3:17*f*, 3:44, 3:68

Consumption: of biofuel, 1:61; of coal, 1:118; of electricity, 5:48, 5:49*f*; of energy, 1:4*f*; of fossil fuels,

2:xiii; household electricity, **5**:49*f;* natural gas, **1**:83–84
Containment ponds, **1**:112
Contra Costa Community College, **2**:4
Controlled burn program, **5**:87
Controller, **3**:11
Control rods, **1**:142
Convection, **2**:68
Conventional water heaters, **5**:67*f*
Conversion, energy loss from, **1**:7–8
Coolants, **1**:141–42
Copenhagen, **3**:58*f*
Corn crops, **4**:93, **4**:96
Corn ethanol, **4**:92–93
Corn gluten meal, **4**:95
Corn kernels, **4**:94*f*
Cornwall Geothermal Project, **4**:41
Corporate aver fleet efficiency (CAFE), **1**:201, **2**:171, **3**:169, **4**:171, **5**:173
Coso geothermal field, **4**:11–12
Cost: of FCVs, **2**:114; of fuel cells, **2**:97, **2**:121–22; geothermal heat pump effectiveness of, **4**:69; of geothermal power plants, **4**:46; of green roofs, **5**:87; of microhydroelectric power plants, **3**:93–94; of solar energy, **2**:30–31; of solar water heaters, **2**:79, **5**:66–67; of wind energy, **3**:17*f,* **3**:44, **3**:68
Costa Rica, **4**:43–44
Countries: CO_2 emissions of, **5**:5; coal production of, **1**:115–16; energy efficient, **5**:21; geothermal energy interest of, **4**:45; green cities of, **5**:92–94; natural gas production of, **1**:81–82; nuclear energy in, **5**:112; oil-producing, **1**:56; using biomass, **4**:85–88; using geothermal energy, **4**:32*f;* using geothermal heat pumps, **4**:73; using microhydroelectric power plants, **3**:94–95; using tidal power, **3**:109–11; using wave energy, **3**:116–18
Cow manure, **4**:98
CRS. *See* Congressional Research Service
Crude oil, **1**:60*f;* global reserves of, **1**:188*t*–189*t,* **2**:158*t*–159*t,* **3**:156*t*–157*t,* **4**:158*t*–159*t,* **5**:160*t*–161*t;* imports of, **1**:xi–xii, **2**:xi–xii, **3**:xi–xii, **4**:xi–xii, **5**:xi–xii; products from, **1**:40*f;* refineries, **1**:46–47; US imports of, **1**:54–56
Crystalline silicon solar cells, **2**:7–8
CSP. *See* Concentrating solar power
Cubic foot, **1**:18
Currie, Linda, **1**:28–31
Custom Coals International, **1**:122
CUTE. *See* Clean Urban Transport for Europe
Cut-in speeds, **3**:21
CYES. *See* California Youth Energy Services

Daimler AG, **2**:86, **5**:119
Dam gates, **3**:81
Dams: for hydroelectric energy, **3**:78–80; types of, **3**:80
Danube river, **3**:87–88, **3**:88*f*
Darajat, **4**:41
Darfur Refugee Camps, **2**:72
da Rosa, Aldo V., **5**:61
Darrieus wind turbines, **3**:13–14
Database of State Incentives for Renewable Energy (DSIRE), **3**:71
Davenport University, **4**:72
Davy, Humphrey, **1**:198, **2**:168, **3**:166, **4**:168, **5**:170
DC. *See* Direct current
Dearborn, Michigan, **5**:86
Deepwater floating wind turbines, **3**:63
Deepwater Horizon well, **1**:61
Deforestation, **1**:120, **4**:102
Delaware, **3**:37

1: Oil, Natural Gas, Coal, and Nuclear
2: Solar Energy and Hydrogen Fuel Cells
3: Wind Energy, Oceanic Energy, and Hydropower
4: Geothermal and Biomass Energy
5: Energy Efficiency, Conservation, and Sustainability

Delhaize American organization, **5**:105
Dell-Winston School Solar Challenge, **2**:45
Denmark: biomass energy source in, **4**:89; wind energy production of, **3**:56–58, **3**:58*f*, **5**:115
Department of Energy (DOE), **1**:34, **1**:166, **2**:136, **3**:134, **4**:136, **5**:138; energy security and, **5**:127; Energy Star program of, **4**:73; hydrogen research of, **2**:100–101; hydrogen storage research of, **2**:123; job vacancies in, **1**:212, **2**:182, **3**:180, **4**:182, **5**:184; net metering information from, **3**:43; wind energy report of, **3**:17, **3**:24, **3**:71
Department of Energy Office of Fossil Energy, **1**:100
Department of Labor, **1**:212, **2**:182, **3**:180, **4**:182, **5**:184
Deposits, coal, **1**:106
De Saussure, Horace Benedict, **1**:204, **2**:174, **3**:172, **4**:174, **5**:176
Desiccant displacement systems, **5**:29
DeSoto Next Generation Solar Energy Center, **1**:xiii, **2**:xiii, **2**:22, **2**:22*f*, **3**:xiii, **4**:xiii, **5**:xiii
Desuperheater, **4**:69
Diesel, Rudolf, **1**:47, **1**:48, **1**:199, **2**:169, **3**:167, **4**:112, **4**:113, **4**:169, **5**:171
Diesel engines, **4**:112*f*; biodiesel emissions compared to, **4**:116; bus fueled by, **1**:48*f*; emissions of, **4**:116; fuel, **1**:47–48; functioning of, **4**:112–13; hydrogen injection in, **2**:113; vegetable oil operation originally for, **4**:113
Diodati, Jason, **1**:49–52
Direct current (DC), **2**:8, **3**:39
Directional drilling, **1**:44–45
Direct Methanol Fuel Cells (DMFCs), **2**:89, **2**:120*f*, **5**:118*f*
Discover Solar Energy, **2**:33, **2**:56
DiscoverThis, **2**:127, **5**:61
Dissolved oxygen levels, **3**:96
Diversion power plant, **3**:81
Dixon, Patrick, **4**:55
DMFCs. *See* Direct methanol fuel cells
DOE. *See* Department of Energy
Dongtan, China, **5**:94
Doping process, **2**:7
Double-flash power plants, **4**:11
Double-pane windows, **5**:52*f*
Dr. FuelCell Science kit, **2**:126
Drake, Edwin L., **1**:67, **1**:204, **2**:174, **3**:172, **4**:174, **5**:176
Drilling: horizontal and directional, **1**:44–45; rig, **1**:44*f*
Driving habits, **5**:75–76
Dry-milling plants, **4**:93
Dry-milling process, **4**:94–95, **4**:94*f*
Dry steam powered plants, **4**:8–9, **4**:8*f*
DSIRE. *See* Database of State Incentives for Renewable Energy
Ducks Unlimited National Headquarters, **5**:87
DuPont, **2**:102
Durability, **4**:70
DVD players, **5**:54
Dye-sensitized solar cells, **2**:11–12, **2**:12*f*

Earth: ecosystem of, **5**:99–100; geothermal heat pump and, **4**:57;

global warming of, **5**:4–5; global winds of, **3**:3*f;* Honor the, **2**:8; human demands on, **5**:7; interior heat energy of, **4**:2–5, **4**:3*f;* science, **1**:218, **2**:188, **3**:186, **4**:188, **5**:190; temperatures and pressures of, **4**:3; temperature underground of, **4**:4, **4**:57; thermal energy of, **4**:15

Eastern Interconnection, **5**:119

Eastport, Maine, **3**:108

Eco-friendly certification, **5**:35

Eco-friendly materials, **5**:8*f*

Eco-friendly schools, **4**:65*f*

Ecological architecture, **5**:24

Ecological footprint, **5**:7

Econar, **4**:75

Economics: biomass benefits of, **4**:104; of geothermal heat pump, **4**:73–74; of solar energy, **2**:30; of solar water heaters, **2**:78; of tidal power, **3**:111; of wind energy, **3**:16–18

Economic stimulus Bill, **1**:33

Economy, hydrogen, **2**:101

Eco-Roof Incentive Programs, **5**:95

Eco-roofs, **5**:65

Eco-structure, in Florida, **5**:96

Ecosystem, of earth, **5**:99–100

Ecoversity, **4**:127

ECR Industries, **4**:75

Edison, Thomas, **1**:199, **2**:1, **2**:169, **3**:167, **4**:169, **5**:171

Edison Electric Co., **1**:198, **2**:168, **3**:166, **4**:168, **5**:170

EERE. *See* Energy Efficiency and Renewable Energy

EFRC. *See* Energy Frontier Research Centers

Egypt, **1**:91, **1**:93*f*

E.I. Dupont De Nemours, **1**:214, **2**:184, **3**:182, **4**:184, **5**:186

Einstein, Albert, **1**:199, **1**:204, **2**:169, **2**:174, **3**:167, **3**:172, **4**:169, **4**:174, **5**:171, **5**:176

Electrical grid system: grid-connected system and, **3**:42–43; main power grids in, **5**:119; residential system connection to, **3**:42–43, **3**:42*f;* of US, **2**:29; US improvement needed in, **5**:119. *See also* Transmission grid; Utility grid

Electricity: biomass' capacity of, **4**:104; CO_2 emission reduction and, **5**:48–50; coal-fired power station producing, **1**:113–14, **1**:114*f;* energy and, **1**:16–17; geothermal energy generating, **4**:8–13; home's requirements of, **3**:16, **3**:41; home's use of, **3**:44, **3**:90, **5**:48, **5**:49*f;* household consumption of, **5**:49*f;* from hydroelectric power plants, **3**:83; kWh of, **3**:15; light producing, **2**:2; from photovoltaic power plants, **2**:16*f;* solar cells producing, **2**:7*f;* solar energy creating, **2**:14–15; sources producing, **1**:16–17; stationary bicycles generating, **5**:9; thin-film solar cells production of, **2**:10; transmission of, **1**:17; US infrastructure modernization for, **3**:68–69; wind energy generating, **3**:38; wind turbines generating, **3**:16, **3**:21–22

Electric motor, **1**:198, **2**:168, **3**:166, **4**:168, **5**:170

Electric Power Research Institute, **3**:110

Electric power transmission system, **3**:25

Electric street cars, **5**:89*f*

Electric Vehicle Association of America, **1**:165, **2**:135, **3**:133, **4**:135, **5**:77, **5**:126, **5**:137

Electric vehicles, **5**:18*f,* **5**:71–74; benefits and challenges of, **5**:73–74; charging station nozzle for, **5**:73*f;* gas powered vehicle conversion

1: Oil, Natural Gas, Coal, and Nuclear
2: Solar Energy and Hydrogen Fuel Cells
3: Wind Energy, Oceanic Energy, and Hydropower
4: Geothermal and Biomass Energy
5: Energy Efficiency, Conservation, and Sustainability

to, **5:**125–26; green, **5:**71–74; hydrogen fuel cell with, **2:**86*f;* infrastructure needed for, **5:**72–73; miles per gallon estimation of, **5:**74; Nissan Leaf as, **5:**74; Saturn EV-1, **5:**72
Electrolysis, **2:**87, **2:**91–92, **2:**92*f*
Electrolyte, **2:**88
Electromagnetic energy, **1:**6
Electrons, **2:**92
El Paso Solar Pond, **2:**81
El Salvador, **4:**42–43
Emissions: biodiesel v. diesel, **4:**116; carbon footprint tracking of, **5:**18; CO_2, **1:**95*f,* **1:**119–20, **1:**120*f;* CO_2 increasing, **5:**4–5; CO_2 reduction of, **5:**48–50; coal-fired plants from, **1:**122, **5:**124; coal-generating plants with, **1:**121; of fossil fuels, **1:**22*f;* fuel and CO_2, **1:**95*f;* of greenhouse gases, **4:**69; greenhouse gases reduction of, **2:**78, **5:**16; natural gas, **1:**90–94; nitrogen oxide, **1:**98; petroleum diesel, **4:**116; sulfur dioxide, **1:**119; sulfur oxide, **4:**117; US CO_2, **1:**120*f;* zero, **1:**122, **5:**73–74, **5:**94. *See also* Carbon dioxide
Empire State Building, **5:**82
Enel, **4:**20
Energized Learning, **5:**22
Energy: careers in, **2:**54; China and India's use of, **5:**107; consumption of, **1:**4*f;* conversion loss of, **1:**7–8;

electricity and, **1:**16–17; forms of, **1:**5–6; fossil fuels for, **1:**8*f,* **1:**24, **1:**36–37; future of, **5:**129; global consumption of, **1:**18–19; global role of, **1:**1–2, **3:**vii–viii; homes saving, **5:**47–48; impact of, **1:**vii–viii, **2:**vii–viii, **3:**vii–viii, **4:**vii–viii, **5:**vii–viii; industries using a lot of, **1:**20–21; Information Administration, **3:**66, **4:**54; landscaping saving, **5:**56–57; law of conservation of, **1:**6–7; manufacturers in, **1:**213–16, **2:**183–86, **3:**181–84, **4:**183–86, **5:**185–88; measuring, **1:**18; nonrenewable sources of, **1:**9–11; policies, **1:**viii, **1:**2–3, **2:**viii, **3:**viii, **4:**viii, **5:**viii; product development in, **1:**213–16, **2:**183–86, **3:**181–84, **4:**183–86, **5:**185–88; renewable energy and, **1:**180*t*–183*t,* **2:**150*t*–153*t,* **3:**148*t*–151*t,* **4:**150*t*–153*t,* **5:**152*t*–155*t;* renewable sources of, **1:**12–16; rotor blades involving, **3:**7–9; Savers, **4:**54, **4:**76; security, **5:**127; sources of, **1:**8–15; storage, **3:**69; time line of, **1:**197–201, **2:**167–71, **3:**165–69, **4:**167–71, **5:**169–73; US history of, **1:**3–4, **1:**4*t;* US supply of, **1:**14*f;* worldwide uses of, **1:**19–21
Energy audits, **5:**20, **5:**24, **5:**54–55
Energy Clean Cities Program, **4:**118
Energy conservation, **1:**25–26, **5:**107; defining, **5:**3; reading materials on, **1:**162–63, **2:**132–33, **3:**130–31, **4:**132–33, **5:**134–35; in schools, **1:**26–28
Energy consumption: Canada's per capita, **1:**22; environmental issues and, **5:**102; future of, **1:**21–22; global, **1:**20*f;* global and regional, **1:**184*t*–187*t,* **2:**154*t*–157*t,*

3:152*t*–155*t*, 4:154*t*–157*t*, 5:156*t*–159*t*; lighting strategies reducing, 1:27; renewable energy production and, 1:180*t*–183*t*, 2:150*t*–153*t*, 3:148*t*–151*t*, 4:150*t*–153*t*, 5:152*t*–155*t*; by sector, 1:174*t*–177*t*, 2:144*t*–147*t*, 3:142*t*–145*t*, 4:144*t*–147*t*, 5:146*t*–149*t*; by source, 1:170*t*–173*t*, 2:140*t*–143*t*, 3:138*t*–141*t*, 4:140*t*–143*t*, 5:142*t*–145*t*; of US, 1:19*f*

Energy efficiency, 1:25–26, 5:12; audits, 1:31; countries, 5:21; defining, 5:3–4; dome homes and, 5:48; in go-green program, 5:3–4; green buildings for, 5:80–81; home heating and, 5:50; of homes, 5:51–54; home's outdoor landscaping and, 5:56–57; reading materials on, 1:162–63, 2:132–33, 3:130–31, 4:132–33, 5:134–35; school programs for, 5:45; in schools, 1:26–28; sustainable development and, 5:107; windows and, 5:33

Energy Efficiency and Renewable Energy (EERE), 1:36, 2:104, 4:76, 4:105, 5:22, 5:128

Energy Frontier Research Centers (EFRC), 1:33

EnergyGuide, 5:50*f*

Energy Independence and Security Act of 2007, 1:201, 2:171, 3:169, 4:171, 5:173

Energy Policy Act of 2005, 1:201, 2:171, 3:169, 4:171, 5:173

Energy Star, 4:73, 5:31, 5:50, 5:53*f*, 5:55

Enhanced geothermal system, 4:12–13, 4:41, 4:46–47, 4:47*f*, 4:48*f*

Enhanced oil recovery (EOR), 1:45

Environmental Energy Technologies Division, 5:22

Environmental issues: climate change and, 1:22–24; of coal, 1:109–11, 1:128–29; of coal-fired power station, 1:119–20; of CSP, 2:50–51; design projects responsible to, 5:31–32; energy consumption and, 5:102; gasification with, 1:127; of geothermal energy, 4:53; of mining, 1:109–11; musicians conscious of, 5:6–8, 5:8*f*; petroleum and, 1:60–62; surface mining and, 1:108; in tidal power, 3:112–13

Environmentalists, 2:29

Environmental Protection Agency (EPA), 1:97, 1:166, 2:136, 3:134, 4:136, 5:138; carbon footprint reduction tips from, 5:19; electric car mileage from, 5:74; Energy Star program of, 4:73; geothermal heat pump efficiency and, 4:57; *Green Vehicle Guide* from, 5:77; human health mission of, 4:76; personal carbon footprint estimation from, 5:20; SmartWay designation of, 5:75; *Space Conditioning: The Next Frontier* by, 4:69

EOR. *See* Enhanced oil recovery

EPA. *See* Environmental Protection Agency

Equinox Fuel Cell SUV, 5:70

EREC. *See* European Renewable Energy Council

Ericsson, John, 1:198, 1:205, 2:5, 2:168, 2:175, 3:166, 3:173, 4:168, 4:175, 5:170, 5:177

Erren, Rudolf, 1:205, 2:175, 3:173, 4:175, 5:177

Erren engines, 1:205

ESHA. *See* European Small Hydropower Association

Ethane, 1:74–75

Ethanol, 4:82*f*, 4:92; benefits of, 4:96; bio, 4:91–92; concerns about, 4:96; corn, 4:92–93; corn crops for, 4:96;

1: Oil, Natural Gas, Coal, and Nuclear
2: Solar Energy and Hydrogen Fuel Cells
3: Wind Energy, Oceanic Energy, and Hydropower
4: Geothermal and Biomass Energy
5: Energy Efficiency, Conservation, and Sustainability

gasoline with, **4**:93*f*, **4**:96; production of, **4**:93–95; wet-milling process in, **4**:95*f*. *See also* Methanol
Ethylene, **1**:53
Europe: air-to-water heat pumps in, **4**:74; geothermal heat pumps in, **4**:72–73; hot dry rock project of, **4**:48–49; hydrogen fuel cell research of, **2**:112–13; wind energy in, **1**:xv–xvi, **2**:xv–xvi, **3**:xv–xvi, **3**:57, **4**:xv–xvi, **5**:xv–xvi
European Association for Battery, Hybrid and Fuel Cell Electric Vehicles, **5**:77
European Renewable Energy Council (EREC), **3**:98
European Small Hydropower Association (ESHA), **3**:98
European Union, **4**:72
European Wind Energy Association, **3**:59
Evaporator coil, **4**:59
Exide Technologies, **5**:2
Experimental aircraft, **2**:106*f*
Experimental vehicle team, **2**:44–45
Exporters, of coal, **1**:116–17, **1**:117*t*
Exxon Mobil, **1**:215, **2**:185, **3**:183, **4**:185, **5**:187

Faraday, Michael, **1**:198, **1**:205, **2**:168, **2**:175, **3**:166, **3**:173, **4**:168, **4**:175, **5**:170, **5**:177
Farmers, of wind energy, **3**:38, **3**:39
Fast breeder reactors, **1**:149–50
Fast neutron reactors, **1**:149–50
Fat to Fuel, **4**:126
Faya, Antnio, **3**:87
FCHV. *See* Fuel-cell hybrid vehicle
FCO. *See* Fuel Cell Quadracycle
FCVs. *See* Fuel cell vehicles
Federal-Aid Highway Act, **1**:200, **2**:170, **3**:168, **4**:170, **5**:172
Fenton Wind Farm, **3**:34
Ferguson, Charles, **1**:154–56, **1**:155*f*
Fermentation, **2**:92, **4**:94–95, **4**:100
Fermi, Enrico, **1**:134, **1**:200, **1**:205, **2**:170, **2**:175, **3**:168, **3**:173, **4**:170, **4**:175, **5**:172, **5**:177
FGD. *See* Flue gas desulphurization
Finland, **4**:88
First Solar, Inc, **2**:22
Fish farms, **4**:17, **4**:19, **4**:50, **4**:51*f*
Fish ladders, **3**:97*f*
Fission reaction, **1**:138
Flash-steam power plants, **4**:10–12, **4**:10*f*
Flat-plate collector, **2**:76
Fleet vehicles, **1**:85–86, **1**:90
Floating nuclear power plants, **1**:151–52
FloDesign Wind turbines, **3**:56
Flores, Jordan, **1**:30
Florida: eco-structure in, **5**:96; LEED certified school in, **5**:33
Florida Solar Energy Center, **2**:11, **2**:33
Flue gas desulphurization (FGD), **1**:111
Fluidized bed combustion systems, **1**:121–22
Food Lion, **5**:105
Ford, Henry, **1**:199, **2**:169, **3**:167, **4**:169, **5**:171
Ford "999," **2**:109
Ford Motor Co., **1**:200, **1**:216, **2**:170, **2**:186, **3**:168, **3**:184, **4**:170, **4**:186, **5**:72, **5**:86, **5**:172, **5**:188
Forebay, **3**:92*f*

Forklifts, **2:**113
Formula 3 racing car, **4:**108*f*
Fort Atkinson School District, **4:**17, **4:**66–67
Fortman, Mark, **5:**10*f*
Fossil Energy Study Guides and Activities, **5:**128
Fossil fuels, **1:**9–19; coal as, **1:**106; consumption of, **2:**xiii; emission levels of, **1:**22*f*; for energy, **1:**8*f*, **1:**24, **1:**36–37; natural gas cleanest of, **1:**94; petroleum as, **1:**41; reading materials on, **1:**159–60, **2:**129–30, **3:**127–28, **4:**129–30, **5:**131–32
Fourneyron, Benoit, **3:**7
Fox River, Wisconsin, **3:**78
France: Chaudes-Aigues, **4:**32; geothermal district heating facilities in, **4:**50; natural gas and, **1:**84; nuclear energy in, **1:**137–38; tidal power energy in, **1:**xvi–xvii, **2:**xvi–xvii, **3:**xvi–xvii, **3:**109, **3:**109*f*, **4:**xvi–xvii, **5:**xvi–xvii
France, Brian, **5:**1–2, **5:**2*f*
Francis, James, **3:**82
Francis reaction turbines, **3:**92, **3:**94
Fraunhofer Institute for Solar Energy Systems, **1:**215, **2:**185, **3:**183, **4:**185, **5:**187
Frazer, Susan, **4:**98–102
Freedom CAR (Cooperative Automotive Research) Program, **2:**108
Freons, **1:**74
Fresnel Stirling engine, **2:**42
Frisch, Otto, **1:**134
Fritts, Charles, **1:**198, **1:**205, **2:**168, **2:**175, **3:**166, **3:**173, **4:**168, **4:**175, **5:**170, **5:**177
Fruit, solar cells from, **2:**13
Fuel: appliances and use of, **1:**178*t*–179*t*, **2:**148*t*–149*t*, **3:**146*t*–147*t*, **4:**148*t*–149*t*, **5:**150*t*–151*t*; CO_2 emissions of, **1:**95*f*; vegetable oils as, **4:**114; world use of, **1:**8*f*
Fuel Cell 2000, **2:**103
Fuel-cell hybrid vehicle (FCHV), **2:**109
Fuel Cell Quadracycle (FCO), **2:**99
Fuel cells, **2:**116; basic applications of, **2:**103; batteries used with, **2:**97; benefits of, **2:**121; Bloom Energy, **5:**121–22; concerns about, **2:**121–23; cost of, **2:**97, **2:**121–22; defining, **2:**87; drawbacks of, **2:**89–90; functioning of, **2:**88, **2:**88*f*; growth of, **2:**106–7; history notes of, **5:**120; home applications of, **2:**117–19; home installation of, **2:**117–19, **2:**118*f*; hydrogen model cars with, **2:**127; Molten Carbonate, **2:**119; on-site, **5:**119; phosphoric acid, **2:**119; production, **2:**104; school education on, **2:**125–26; small, **2:**106, **2:**120, **2:**120*f*; solid oxide, **2:**107, **5:**121; space shuttles using, **5:**120; stationary systems of, **2:**119; telecommunications using, **2:**119–20; transportation applications of, **2:**107–16; types of, **2:**89. *See also* Hydrogen fuel cells
Fuel Cell Technologies Program, **2:**104
Fuel Cell Test and Evaluation Center, **2:**98
Fuel cell vehicles (FCVs), **2:**89, **2:**98, **2:**99, **2:**107, **2:**108*f*, **5:**69–70; cost of, **2:**114; in Germany, **2:**110–11; in Japan, **2:**109; refueling, **2:**114–15; in United Kingdom, **2:**110; in US, **2:**107–9
Fuel economy, **1:**51
Fuel rods, **1:**140–41, **1:**141*f*
Fuelwood, **1:**14, **2:**71–72, **4:**102
Fuller, Buckminster, **1:**205, **2:**175, **3:**173, **4:**175, **5:**177

1: Oil, Natural Gas, Coal, and Nuclear
2: Solar Energy and Hydrogen Fuel Cells
3: Wind Energy, Oceanic Energy, and Hydropower
4: Geothermal and Biomass Energy
5: Energy Efficiency, Conservation, and Sustainability

Fuller, Calvin, **1**:205, **2**:5, **2**:175, **3**:173, **4**:175, **5**:177
Fumaroles, **4**:52
Fundamentals of Renewable Energy Processes (de Rosa), **5**:61
Furling, **3**:8
FutureGen, **1**:123

Gabcikovo Dam, **3**:98
Gap headquarters, **5**:83
Garden roofs, **5**:65, **5**:78, **5**:84
Garner, Mark, **3**:74, **5**:108, **5**:108*f*
Garst, Charlotte, **3**:65
Gas, **1**:45; injection, **1**:45; offshore drilling for, **1**:80*f*; production, **1**:62
Gas-fired turbine, **3**:32
Gasholder, **4**:98
Gasification, **1**:123; biomass plant for, **4**:79; coal, **1**:125–28, **1**:126*f*; environmental issues with, **1**:127; of wood, **4**:97
Gasohol, **4**:92
Gasoline, **1**:41, **1**:47, **4**:93*f*, **4**:96
Gasoline gallon equivalent (GGEs), **1**:91
Gasper, Peter, **3**:87
Gas powered vehicle conversion, **5**:125–26
GCHP. *See* Closed-loop ground-coupled heat pump
Gearbox, of wind turbines, **3**:9–11
General Electric, **1**:145, **1**:215, **2**:185, **3**:183, **4**:185, **5**:187
General Motors, **2**:108*f*, **2**:114–15; Chevy Volt from, **5**:71; Equinox Fuel Cell SUV of, **5**:70; hydrogen research and development by, **2**:109; Saturn division of, **5**:71; Saturn EV-1 electric car of, **5**:72, **5**:122
Generators: building wind, **3**:51; of hydroelectric plant, **3**:82–83; of microhydroelectric power plant, **3**:90; of wind turbine, **3**:9–10
Geo-Heat Center, **4**:76
Geological Survey, US, **1**:67
Geology, **4**:63
GEO Mission, **4**:54
GeoThermal, **4**:54
Geothermal Education Office, **4**:29
Geothermal energy, **1**:13; air quality standards and, **4**:25; Alaska's resources of, **4**:18–19; for aquaculture, **4**:50; in Arizona, **4**:18; benefits of, **4**:2, **4**:52; binary plant of, **4**:40*f*; in California, **4**:14–15; in Canada, **4**:38–39; Chevron largest producer of, **4**:41; China's resources in, **4**:39; CO_2 credits from, **4**:51; in Colorado, **4**:17, **4**:51*f*; in Costa Rica, **4**:43–44; countries interested in, **4**:45; countries using, **4**:32*f*; defining, **4**:2–5; electricity generated from, **4**:8–13; in El Salvador, **4**:42–43; environmental issues of, **4**:53; Fort Atkinson School District using, **4**:17; France's district heating facilities with, **4**:50; future of, **4**:27–28, **4**:53; Germany's resources in, **4**:45; greenhouses heated by, **4**:23, **4**:50; harnessing, **4**:7; in Hawaii, **4**:16–17; heating system with, **4**:22–23, **4**:50; history of, **4**:5–6; in Iceland, **1**:197, **2**:167, **3**:165, **4**:37–38, **4**:167, **5**:169; in Italy, **4**:35–37; in Japan, **4**:35,

4:50–51; Kalina system and, 4:49; Kenya's resources of, 4:43; land use and, 4:24; locations of, 4:6–7, 4:29; in Mexico, 4:34–35; Minnesota using, 4:66; in Montana, 4:20; in national parks, 4:52; in Nevada, 4:15; in New Mexico, 4:17–18; new technologies employed in, 4:18; New Zealand's resources in, 4:44; Oregon Institute of Technology using, 4:1–2; in Philippines, 4:32–34; reading materials on, 1:162, 2:132, 3:130, 4:132, 5:134; South Dakota's schools using, 4:67; in sustainable development, 5:109–10; tax revenue from, 4:27; in Thailand, 4:39–40; Turkey's resources of, 4:41; in US, 1:xvii, 2:xvii, 3:xvii, 4:xvii, 4:13–20, 4:31, 5:xvii, 5:110f; US companies in, 4:20; used in Toledo Zoo, 4:67; uses for, 4:49–51; in Utah, 4:19–20; Williston Northampton School and, 4:66f; Wisconsin using, 4:66–67

Geothermal Energy Association, 1:35, 4:27–28

Geothermal fluids, 4:9f

Geothermal heat pumps: advantages of, 4:74–75; Arizona testing, 4:65–66; benefits of, 4:68–69; as closed-loop systems, 4:60–61, 4:61f; CO_2 reduced by, 4:68, 4:69; cost-effectiveness of, 4:69; countries using, 4:73; desuperheater used with, 4:69; disadvantages of, 4:75; durability and maintenance of, 4:70; earth's underground temperature used by, 4:57; economics of, 4:73–74; EPA and efficiency of, 4:57; in Europe, 4:72–73; functioning of, 4:59–60, 4:60f; geology required for, 4:63; growth of, 5:109–10; for homes, 5:63; hydrology required for, 4:63; in Idaho, 4:71–72; industry growth of, 4:58–59, 4:58f; in Iowa Central Community College, 4:71; in Kentucky, 4:65, 4:70–71; land use and, 4:63–64; manufacturers of, 4:75; Massachusetts' schools using, 4:67–68; in Michigan, 4:72; in Mississippi, 4:72; Nebraska's schools using, 4:68; in North Dakota, 4:72; as open-loop systems, 4:61–63, 4:62f; radial drilling with, 4:59, 4:77; school benefits of, 4:64; schools using, 4:64–68; site evaluation for, 4:63–64; in Sweden, 4:72; in US, 4:59, 4:70–73; US installed capacity of, 4:70; water heaters using, 4:62–63

Geothermal power plants: advantages of, 4:24–25; in Australia, 4:42; binary, 4:22; Birdsville, 4:42; cost factors of, 4:46; at the Geysers, 4:14f; history of, 4:55; in Iceland, 4:26; in Idaho, 4:15–16; in Indonesia, 4:40; in Larderello, 4:36f; in New Zealand, 4:44f; ORC, 4:21f, 4:24f; single-flash, 4:11; in United Kingdom, 4:41–42

Geothermal reservoir, 4:10, 4:22

Geothermal Resources Council, 4:6, 4:29

Geothermal Steam Act Amendments, 4:27

Geothermal turbines, 4:35

Geothermal wells, 4:51f, 4:65f, 4:66f, 4:67, 4:68, 5:104

Gerdeman, Frederick, 1:206, 2:176, 3:174, 4:176, 5:178

Germany: clean coal technology in, 1:124; coal plants removal in, 5:106; coal production of, 1:115–16; FCV's in, 2:110–11; geothermal resources in, 4:45;

> 1: Oil, Natural Gas, Coal, and Nuclear
> 2: Solar Energy and Hydrogen Fuel Cells
> 3: Wind Energy, Oceanic Energy, and Hydropower
> 4: Geothermal and Biomass Energy
> 5: Energy Efficiency, Conservation, and Sustainability

green roofs in, **5**:83–84; hydrogen fuel cells in, **2**:117; natural gas consumption of, **1**:83–84; photovoltaic systems in, **2**:24–25; solar panels in, **2**:27*f*; wind farms in, **3**:53, **3**:57

Geysers, **4**:4–5, **4**:52

The Geysers, **4**:5–6, **4**:8–9, **4**:14, **4**:14*f*

GGEs. *See* Gasoline gallon equivalent

Gill, Dena, **3**:65

Ginori Conti, Piero, **4**:36*f*

Global Biofuels Outlook: 2009–2015, **5**:120

Global consumption: of coal, **1**:103; of energy, **1**:18–19, **1**:20*f*, **1**:184–87*t*, **2**:154–57*t*, **3**:152–55*t*, **4**:154–57*t*, **5**:156–59*t*; of natural gas, **1**:72–73, **1**:94–95; wind energy and, **3**:56

Global economy: energy driving, **1**:1–2; energy's role in, **3**:vii–viii

Global emissions, of CO_2, **1**:192–95, **2**:162–65, **3**:160–63, **4**:162–65, **5**:5, **5**:105, **5**:164–67

Global installations, **2**:78–79

Global leaders, **2**:24–26

Global Learning, Inc, **2**:33, **5**:127

Global reserves: of crude oil, **1**:188*t*–189*t*, **2**:158*t*–159*t*, **3**:156*t*–157*t*, **4**:158*t*–159*t*, **5**:160*t*–161*t*; of natural gas, **1**:79–81, **1**:79*f*, **1**:188*t*–189*t*, **2**:158*t*–159*t*, **3**:156*t*–157*t*, **4**:158*t*–159*t*, **5**:160*t*–161*t*

Global warming, **1**:viii, **1**:23–24, **1**:25–33, **1**:94, **2**:viii, **3**:viii, **4**:viii, **5**:viii, **5**:106*f*; CO_2 increasing causing, **5**:4–5; synthetic natural gas and, **1**:128

Global winds, **3**:3*f*

Glycerin, **4**:112

Go-green program: energy efficiency in, **5**:3–4; of NASCAR, **5**:1–3

Go-green project: in California, **5**:33–34; Empire State Building and, **5**:82; in Kentucky, **5**:32

Go-green public schools, **5**:31–36

Golden Gate Bridge, **3**:111, **3**:112*f*

Gomez, Cesar, **1**:50

Goodyear Tire & Rubber Co., **5**:2

Gore, Al, **5**:129

Government: organization websites and, **1**:165–67, **2**:135–37, **3**:133–35, **4**:135–37, **5**:137–39; US, **4**:80

Gradient zones, **2**:81

Grand Coulee Dam, **3**:75, **3**:76*f*

Granite quarry, **4**:64

Grant, John D., **1**:206, **2**:176, **3**:174, **4**:176, **5**:178

Grätzel, Michael, **2**:11

Gravity dam, **3**:80

Gray, Rande, **5**:102–5

Greasecar, **1**:51, **4**:124

Great Geysir, **4**:37

Great Rift Valley, **4**:43

Great Seneca Creek Elementary School, **5**:35

Green architecture, **5**:88–89

Green biz, **1**:210, **2**:180, **3**:178, **4**:180, **5**:182

Green Building Initiatives, **5**:95

Green Building Rating System, **5**:26–27

Green buildings: components of, **5**:79; construction materials for, **5**:80–81; energy efficiency of, **5**:80–81; natural environment and, **5**:80

Green careers guide, **1:**210, **2:**180, **3:**178, **4:**180, **5:**182
Green certification, **5:**27*f*
Green cities: in other countries, **5:**92–94; ranking criteria of, **5:**90; snapshots of, **5:**90–92; in US, **5:**89–92, **5:**92*t*
Green Club, **2:**19–20
Green Cluster, **5:**77
Green-Collar Jobs report, **2:**80
Green community, **4:**65*f*
Green corps, **1:**211, **2:**181, **3:**179, **4:**181, **5:**183
Green-e, **3:**97
Green energy jobs, **1:**211, **2:**181, **3:**179, **4:**181, **5:**183
Green Existing Tool Kit, **5:**41
Green Faith in Action Project, **1:**29, **1:**30
Green Grid trays, **5:**40
Greenhouse, geothermal heated, **4:**23, **4:**50
Greenhouse effect, **2:**62, **2:**62*f*, **2:**66, **2:**77*f*, **5:**4
Greenhouse gases, **1:**22–24, **5:**106*f*; CO_2 as, **1:**23; emission reduction of, **2:**78, **5:**16; emissions of, **4:**69; human caused, **5:**4*f*; law, **3:**32; natural gas and, **1:**94; nuclear energy and, **1:**11; solar water heaters reducing, **2:**78; US emissions reduction target for, **5:**16
GreenLearning Canada, **5:**127
Greenpeace, **1:**36
Green Power Network Net Metering, **3:**71
Green Roof Construction and Maintenance (Luckett), **5:**40
Green Roof for Healthy Cities, **5:**42
Green Roof Plants (Snodgrass and Snodgrass), **5:**40
Green roofs, **1:**26*f*, **5:**42–43, **5:**82; architecture for, **5:**81–82; around the world, **5:**83–84; benefits of, **5:**86; components of, **5:**84–86, **5:**85*f*; cost of, **5:**87; in Dearborn, Michigan, **5:**86; effectiveness of, **5:**84; in Germany, **5:**83–84; of Hanneford Supermarket, **5:**103; for homes, **5:**65–66, **5:**65*f*; issues with, **5:**87; L'Historial de la Vendée with, **5:**83, **5:**83*f*; maintenance of, **5:**86; of school buildings, **5:**38–40, **5:**38*f*
Greenroofs.com, **5:**95
Greensburg, Kansas, **5:**47
Greensburg tornado, **5:**45
Greensburg Wind Farm, **3:**34
Green School Buildings, **5:**42
Green Schools Program, **1:**27–28
Green Vehicle Guide, **5:**77
Green vehicles, **5:**68–76; car maintenance and, **5:**75–76; driving habits and, **5:**75–76; electric, **5:**71–74
Green Vision program, **5:**90
GreenWood Resources, **4:**84*f*
Grid-connected system, **3:**42–43, **3:**42*f*
Grieves, Tim, **3:**19–23
Groundwater, **4:**26–27
Grove, William, **1:**206, **2:**176, **3:**174, **4:**176, **5:**120, **5:**178
Guatemala, **4:**34
Guiding Stars, **5:**104
Gulf of Mexico, **1:**61
Gunung Salak, **4:**41
Guorui, Luo, **4:**103
Guri Dam, **1:**xvii, **2:**xvii, **3:**xvii, **4:**xvii, **5:**xvii
Guri Hydroelectric power plants, **3:**87
Gutierrez, Maricruz, **1:**50

H_2SO_4. *See* Sulfuric acid
Hahn, Otto, **1:**134
Halliday, Daniel, **1:**206, **2:**176, **3:**174, **4:**176, **5:**178
Hancock County Wind Energy Center, **3:**34

1: Oil, Natural Gas, Coal, and Nuclear
2: Solar Energy and Hydrogen Fuel Cells
3: Wind Energy, Oceanic Energy, and Hydropower
4: Geothermal and Biomass Energy
5: Energy Efficiency, Conservation, and Sustainability

Hanneford Supermarket, **5:**103–4, **5:**103*f*
Harman, Stephanie, **2:**62–66, **2:**63*f*
Harriman, Chris, **4:**16*f*
Harris, Matt, **2:**8
Hashimoto, Ryutaro, **5:**106*f*
Hawaii, **4:**16–17
the Head, **3:**91–92
Heat, **1:**6–8
Heating system, **4:**22–23, **4:**50, **5:**25
Heat pumps, **1:**13
Heifer International, **5:**91, **5:**91*f*
Heliocentris Solar Hydrogen Fuel Cell kit, **2:**125
Heliostats, **2:**47, **2:**48*f*
Henry Sibley Senior High School, **5:**10*f*
Herbert Bryant Conference Center, **4:**72
Herschel, John, **2:**5
HFC. *See* Hydrogen fuel cells
High-level nuclear waste, **1:**145
High-level radioactive wastes (HLRW), **1:**34
High-oil algae, **2:**93
High pressure, **3:**4
High Winds Energy Center, **3:**32–33
Hilderbrand, John, **3:**39
HLRW. *See* High-level radioactive wastes
Hobby Lobby, **5:**61
Home entertainment systems, **5:**54
Homes: appliances in, **5:**52–53; battery storage for, **2:**15*f*; biodiesel heating of, **4:**120; blower door test of, **5:**56*f*; carbon footprints of, **5:**8–9; concrete dome, **5:**46, **5:**47*f*; efficient heating of, **5:**50; electricity requirements of, **3:**16, **3:**41; electricity use of, **3:**44, **3:**90, **5:**48, **5:**49*f*; energy audits of, **5:**54–55; energy efficiency dome, **5:**48; energy efficiency of, **5:**51–54, **5:**56–57; energy saving, **5:**47–48; fuel cell applications at, **2:**117–19; fuel cell installation in, **2:**117–19, **2:**118*f*; fuels and appliances used in, **1:**178*t*–179*t*, **2:**148*t*–149*t*, **3:**146*t*–147*t*, **4:**148*t*–149*t*, **5:**150*t*–151*t*; of future, **5:**46–47; geothermal heat pumps for, **5:**63; green roofs for, **5:**65–66, **5:**65*f*; heating and cooling tips for, **5:**51–52; home entertainment systems in, **5:**54; hydrogen fuel cell applications for, **5:**118–19; landscaping of, **5:**57*f*; lighting systems of, **5:**53; microhydroelectric power plants and, **3:**91*f*; passive solar design of, **2:**68–69; renewable energy for, **5:**63–64; smaller, **5:**55; Solar Decathlon of, **5:**64–65, **5:**64*f*; solar energy for, **5:**63–64; solar water heaters for, **5:**64; Wind Energy for, **3:**26, **3:**72; wind turbines for, **5:**63
Home Solar Panels, **5:**77
Honda FCX Clarity, **2:**86*f*, **2:**109
Hong Kong ferry boats, **2:**6
Honor the Earth (HTE), **2:**8
Hoover Dam, **1:**17*f*, **3:**76–77
Hopi Indians, **1:**104
Horizon Fuel Cell Technologies, **2:**127, **5:**69
Horizontal-axis turbines, **3:**12–13, **3:**13*f*, **3:**40–41
Horizontal drilling, **1:**44–45
Horizontal ground loops, **4:**64

Horrell, J. Scott, **1:**110*f*
Horse Hollow Wind Energy Center, **3:**31, **3:**31*f*
Hot dry rock, **4:**12–13, **4:**12*f*, **4:**41–42, **4:**49; Australia's resources in, **4:**47–48; enhanced geothermal system v., **4:**46–47; European project of, **4:**48–49
Hot rod, Model T, **2:**95–99, **2:**95*f*
Hot Springs, Arkansas, **4:**5
Hot springs, outdoor, **4:**52*f*
HowStuffWorks web site, **1:**68, **2:**55
HTE. *See* Honor the Earth
Human body, carbon in, **5:**7
Human health, **4:**76
Hurricane Katrina, **2:**119
Hybrid cars, **5:**68, **5:**69*t*, **5:**78
Hybrid poplars, **4:**83–85, **4:**84*f*
Hybrid systems, **3:**122
Hydrocarbons, **1:**47, **1:**74–75
Hydroelectric energy, **1:**xvi, **1:**12–13, **2:**xvi, **3:**xvi, **4:**xvi, **5:**xvi; advantages of, **3:**95–96; Canada's generation of, **3:**85–86; dams built for, **3:**78–80; along Danube river, **3:**87–88, **3:**88*f*; defining, **3:**77; disadvantages of, **3:**96–98; history of, **3:**77; in India, **3:**89; kinetic energy from, **3:**73–74; large-scale, **3:**100; in Norway, **1:**xvi, **2:**xvi, **3:**xvi, **3:**84*f*, **3:**87, **4:**xvi, **5:**xvi, **5:**112; reading materials on, **1:**161, **2:**131, **3:**129, **4:**131, **5:**133; in Romania, **3:**88–89; in sustainable development, **5:**108–9, **5:**108*f*; tidal power creating, **3:**104; turbine improvement in, **3:**82; in US, **3:**74–77, **3:**74*f*
Hydroelectric power plants: in Austria, **3:**89; components of, **3:**81–83; electricity from, **3:**83; global leading, **3:**83, **3:**84*f*; Guri, **3:**87; Itaipú, **3:**86–87, **3:**86*f*; in Italy, **2:**94; Simón Bolivar, **3:**87; small-scale, **3:**89–90; types of, **3:**80–81, **3:**82*f*

Hydrogen, **2:**104; algae producing, **2:**92–94; basics of, **2:**87; blue-green algae producing, **2:**94; Canadian highway with, **2:**111; diesel trucks injection of, **2:**113; DOE research on, **2:**100–101; DOE storage research on, **2:**123; economy, **2:**101; fuel cell model cars, **2:**127; gas tanks, **2:**122, **2:**123*f*; General Motors' research and development of, **2:**109; history using, **2:**90; InfoNet, **1:**165, **2:**135, **3:**133, **4:**135, **5:**137; Italy's power plant using, **2:**94; Norway's refueling for, **2:**114, **2:**115*f*; power plants, **2:**94; production of, **2:**91–100; Riversimple car using, **2:**110, **2:**110*f*, **5:**69; storage, **2:**122–23; technology research on, **2:**99–100; temperature and, **2:**87; thermochemical, **2:**92; uses of, **2:**90–91; vehicle, **2:**116
Hydrogen fuel cells (HFC), **1:**xv, **1:**15, **2:**xv, **2:**85–86, **2:**128, **3:**xv, **4:**xv, **5:**xv; aircraft propulsion with, **2:**105; buses using, **2:**111–13, **2:**112*f*; Canada's buses using, **2:**111; companies making, **2:**102; electric vehicle with, **2:**86*f*; Europe's research of, **2:**112–13; experimental aircraft powered by, **2:**106*f*; functioning of, **2:**88*f*, **5:**120; future of, **2:**100–101; in Germany, **2:**117; home applications of, **5:**118–19; in Japan, **2:**118–19; model racing cars with, **2:**101–2; Model T hot rod running on, **2:**95–99, **2:**95*f*; NASA using, **1:**16*f*; reading materials on, **1:**160–61, **2:**130–31, **3:**128–29, **4:**130–31, **5:**132–33; sales growth of, **5:**116; specialty transportation using, **2:**113–14; in sustainable development, **5:**116–19; synthetic natural gas and, **1:**127–28; transportation application of, **5:**117–18;

> 1: Oil, Natural Gas, Coal, and Nuclear
> 2: Solar Energy and Hydrogen Fuel Cells
> 3: Wind Energy, Oceanic Energy, and Hydropower
> 4: Geothermal and Biomass Energy
> 5: Energy Efficiency, Conservation, and Sustainability

US buses using, **2**:111; US cutting funding for, **2**:86. *See also* Fuel cell vehicles
Hydrogenics, **2**:97, **2**:102
Hydrogen sulfide, **1**:75, **4**:25, **4**:26*f*
Hydrology, **4**:63
Hydrophobic nanocoating technologies, **5**:117*f*
Hydropower Program, **3**:124
Hydrothermal fluids, **4**:8*f*, **4**:10*f*
Hyundai Motor Co., **2**:86, **2**:111, **5**:69, **5**:119

IAEA. *See* International Atomic Energy Agency
IBM, **1**:215, **2**:185, **3**:183, **4**:185, **5**:187
Iceland, **2**:112, **4**:38*f*; geothermal energy in, **1**:197, **2**:167, **3**:165, **4**:37–38, **4**:167, **5**:169; geothermal power plants in, **4**:26
ICS. *See* Integral collector-storage systems
Idaho: biodiesel projects of, **4**:118; carbon footprint reduction in, **5**:11–12; geothermal heat pumps in, **4**:71–72; geothermal power plants in, **4**:15–16
Idaho National Laboratory's Geothermal Program, **4**:76
IGCC. *See* Integrated gasification combined cycle
IHA. *See* International Hydropower Association

Illinois, **4**:119–20
Illinois EPA Green School Checklist, **5**:12*f*
Imports: of coal, **1**:117; of crude oil, **1**:xi–xii, **2**:xi–xii, **3**:xi–xii, **4**:xi–xii, **5**:xi–xii
Impoundment hydropower plants, **3**:80, **3**:82*f*
India, **2**:72; coal production of, **1**:116; energy use of, **5**:107; hydroelectric energy in, **3**:89; wind energy in, **3**:59–60
Indiana, **4**:68, **4**:120
Individuals carbon footprint, **5**:6*f*
Indonesia: biomass energy source in, **4**:86; geothermal power plants in, **4**:40; sugarcane field in, **4**:86*f*
Industrial Revolution, **1**:10, **1**:24, **1**:115
Industry: aluminum, **1**:21; energy used by, **1**:20–21; geothermal heat pumps growth and, **4**:58–59, **4**:58*f*; of natural gas, **1**:78*f*; natural gas used in, **1**:72; oil, **3**:16; stationary fuel cell systems in, **2**:119; steel, **1**:20–21; US coal, **1**:113–15, **1**:115*f*, **1**:117; Worldwide Fuel Cell, **2**:106
Infrastructure: for electric vehicles, **5**:72–73; hydrogen requiring, **2**:101; US modernizing of, **3**:68–69
Insulation, **5**:52
Integral collector-storage systems (ICS), **2**:77
Integrated gasification combined cycle (IGCC), **1**:122
Interior heat energy, **4**:2–5, **4**:3*f*
International Association for Natural Gas Vehicles, **1**:100
International Atomic Energy Agency (IAEA), **1**:135–36, **1**:166, **2**:136, **3**:134, **4**:136, **5**:138
International Geothermal Association, **4**:20

International Green Roof Association, **5:**96
International Ground Source Heat Pump Association, **4:**76–77
International Hydropower Association (IHA), **3:**99
International Journal on Hydropower and Dams, **3:**99
International Organization for Standardization (ISO) Technical Committee on Hydrogen Technologies, **2:**124
International Partnership for a Hydrogen Economy (IPHE), **2:**124
International Renewable Energy Agency (IRENA), **5:**93*f*
International Solar Energy Society, **2:**33, **2:**83
Interstate Renewable Energy Coalition, **2:**16
Inverter, **3:**42, **3:**90
Ions, **2:**88
Iowa, **5:**35
Iowa Central Community College, **4:**71
Iowa Stored Energy Park, **3:**70
IPHE. *See* International Partnership for a Hydrogen Economy
IRENA. *See* International Renewable Energy Agency
Iron Gate Dam I, **3:**88, **3:**88*f*
Itaipú hydroelectric power plants, **3:**86–87, **3:**86*f*
Italy: geothermal energy in, **4:**35–37; hydrogen power plant in, **2:**94

James, David, **4:**127
Japan: FCV's in, **2:**109; geothermal energy sources in, **4:**35, **4:**50–51; hydrogen fuel cells in, **2:**118–19; Kirishima City, **4:**52*f*; rooftop garden in, **5:**84; solar energy and, **2:**83; solar powered cargo ships of, **2:**27; solar systems installed in, **2:**25–26

Jiu River, **3:**88–89
John Day Dam, **3:**75–76
JSS. *See* Junior Solar Sprint
Juarez, Andres, **1:**50
Junior Solar Sprint (JSS), **1:**88, **2:**127
Junior Solar Sprint/Hydrogen Fuel Cell (JSS/HFC), **2:**127

K-9 Comfort Cottage, **2:**11
Kaipara Harbor, **3:**111
Kalina system, **4:**49
Kansas, **3:**34, **5:**47
Kazimi, Mujid, **1:**206, **2:**176, **3:**174, **4:**176, **5:**178
Keahole Point, Hawaii, **3:**121, **3:**121*f*
Keighley, Seth, **4:**110*f,* **4:**111
Kelp, **1:**98
Kemp, Clarence, **2:**59
Kentucky: biodiesel school buses in, **4:**119; geothermal heat pumps in, **4:**65, **4:**70–71; go-green projects in, **5:**32
Kenya, **4:**43
Keros, Alex, **2:**108*f*
Kerosene, **1:**53
Kibaki, Mwai, **4:**43
KidWind Project, **3:**47–51
Kilauea Volcano, **4:**16
Kill A Watt, **5:**62
Kilowatt-hour (kWh), **1:**18, **3:**15
Kinetic energy, **1:**5; from hydroelectric energy, **3:**73–74; from oceans, **3:**103–4; temperature and heat as, **1:**6
Kirishima City, Japan, **4:**52*f*
Kirwan, Kerry, **4:**108*f*
Krocker, J.D., **4:**6
kWh. *See* Kilowatt-hour
Kyoto Box oven, **2:**71
Kyoto International Convention, **5:**106
Kyoto Protocol, 1997, **1:**32–33, **1:**201, **2:**171, **3:**169, **4:**171, **5:**16, **5:**105–6, **5:**106*f,* **5:**173

> 1: Oil, Natural Gas, Coal, and Nuclear
> 2: Solar Energy and Hydrogen Fuel Cells
> 3: Wind Energy, Oceanic Energy, and Hydropower
> 4: Geothermal and Biomass Energy
> 5: Energy Efficiency, Conservation, and Sustainability

LaDuke, Winona, **2**:8
Lake County-Southeast Geysers Effluent Pipeline Project, **4**:6
Landfills: biomass gas from, **1**:97; methane gas recovered from, **1**:97–98, **2**:120, **4**:97; nitrogen oxide emissions of, **1**:98
Land of Volcanoes, **4**:37
Landscaping: energy efficiency and, **5**:56–57; of homes, **5**:57*f*; saving energy, **5**:56–57
Land use: CSP plants and, **2**:50; geothermal energy and, **4**:24; geothermal heat pump and, **4**:63–64; solar energy and, **2**:29
La Rance River tidal power plant, **3**:109*f*
Larderello, **4**:36–37, **4**:36*f*
Large-scale hydroelectric energy, **3**:100
Las Pailas Geothermal Plant, **4**:43
Lavoisier, Antoine, **2**:90
Law of conservation of energy, **1**:6–7
Leadership in Energy and Environmental Design (LEED), **5**:2, **5**:26, **5**:31, **5**:32, **5**:80, **5**:82. *See also* LEED certification
LEDs. *See* Light emitting diodes
LEED. *See* Leadership in Energy and Environmental Design
LEED certification: American Federation of Teachers and, **5**:37; Florida school with, **5**:33; plaque of, **5**:81*f*; Pleasant Ridge Montessori School with, **5**:33; schools with, **5**:28–32, **5**:82; of Summerfield Elementary School, **5**:35–36; supermarket with, **5**:102–5
Legislation, on carbon footprint, **5**:21
Lentz, Timothy, **5**:64*f*
Lewis, Mike, **2**:98
Lewis, Zane, **2**:95*f*, **4**:110*f*, **4**:111
Leyte Geothermal Production Field, **4**:33
L'Historial de la Vendée, **5**:83, **5**:83*f*
Life science, **1**:218, **2**:188, **3**:186, **4**:188, **5**:190
Light emitting diodes (LEDs), **5**:53
Lighting strategies, **1**:27
Lighting systems, **5**:53
LIH. *See* Low-impact hydropower facilities
Limestone, **1**:42
Liquefied petroleum gas (LPG), **1**:53, **1**:85
Liquid hydrogen tanks, **2**:123
Liquid metal fast breeder reactors, **1**:150*f*
Liquid refrigerant, **4**:59
Lithium, **5**:124
Little Rock, Arkansas, **5**:91, **5**:91*f*
LLW. *See* Low-level nuclear waste
Local winds, **3**:4
London, England, **5**:93
Long-term storage, of carbon dioxide, **1**:122–23
Lorusso, Jarred, **5**:125–26, **5**:126*f*
Los Alamos National Laboratory, **1**:166, **2**:136, **3**:134, **4**:13, **4**:136, **5**:138
Louisiana, **4**:112
Low-impact hydropower facilities (LIH), **3**:97
Low-level nuclear waste (LLW), **1**:145–46
Low pressure, **3**:4
Low-temperature solar collectors, **1**:12

LPG. *See* Liquefied petroleum
Luckett, Kelly, **5**:40
Luminant, **3**:31
Lund, John W., **4**:20–23

Macapagal-Arroyo, Gloria, **4**:87
Macari Family Foundation, **5**:125
Maggs, Steve, **4**:108*f*
Maine Public Utilities Commission Program, **5**:62
Maintenance: car, **5**:75–76; geothermal heat pump, **4**:70; of green roofs, **5**:86; wind turbines, **3**:41
Mak-Ban, **4**:41
Malaysia, **4**:87
Mammoth Pacific power plant, **4**:10
Manhattan Project, **1**:200, **2**:170, **3**:168, **4**:170, **5**:172
The Manhattan Project, **1**:134
Manufacturers: China's solar cell, **1**:xiv, **1**:201, **2**:xiv, **2**:24, **2**:171, **3**:xiv, **3**:169, **4**:xiv, **4**:171, **5**:xiv, **5**:173; in energy, **1**:213–16, **2**:183–86, **3**:181–84, **4**:183–86, **5**:185–88; geothermal heat pump, **4**:75; nuclear reactors, **1**:145; paper, **1**:21; wind turbines, **3**:16
Mariculture, **3**:123
Marine organisms, **1**:41–42, **3**:123
Marine plants, **1**:98
Marquez, Abigail, **1**:50
Martinez, Xiomara, **1**:50
Maryland, **5**:35
Masdar City, **3**:92–93, **5**:93*f*
Massachusetts: Boston, **5**:91; geothermal heat pump used in, **4**:67–68
Mastaitis, Vicki, **1**:xiii, **2**:xiii, **2**:16, **3**:xiii, **4**:xiii, **5**:xiii
Mayer, John, **5**:8*f*
Mayors Climate Protection Center, **5**:16
Mazda Premacy Hydrogen RE, **5**:70

McCurdy, Ross, **2**:94–99, **2**:125, **2**:126, **4**:110–11
McDonough, William, **5**:86
MCFC. *See* Molten Carbonate fuel cells
McGrath, Gerald, **5**:66–68
Meager Mountain, **4**:38
Medford Township school district, **4**:119*f*
Meitner, Lise, **1**:134
Mendoza, Crystal, **1**:50
Mercedes BlueZero F-Cell, **5**:70, **5**:70*f*
Mercymount Country Day School, **5**:71
Meredith, James, **4**:108*f*
Methane, **1**:74–75, **1**:75*f*, **4**:96; cattle source of, **1**:76; drawbacks of, **4**:101; landfills producing, **1**:97–98, **2**:120, **4**:97; marine plants producing, **1**:98
Methane hydrate, **1**:95–97; deposits of, **1**:96*f*; research needed on, **1**:96–97
Methanol, **2**:89, **2**:120, **2**:120*f*
Methyl esters, **4**:112
Mexico, **4**:34–35
Michigan, **4**:72
Microhydroelectric power plants, **3**:89–90, **3**:91*f*; China using, **3**:94; components of, **3**:90–93, **3**:92*f*; cost of, **3**:93–94; countries using, **3**:94–95; generator of, **3**:90; homes and, **3**:91*f*; Shutol, **3**:95*f*; terrain required for, **3**:93; US potential of, **3**:96, **3**:98
Microsoft Corporation, **5**:18
Microturbines, **2**:41
Middlebury College, **4**:79–80, **4**:80*f*
Middle East, **1**:79*f*
Migratory fish, **3**:87
Miles per gallon estimation, **5**:74
Military, US, **1**:137

> 1: Oil, Natural Gas, Coal, and Nuclear
> 2: Solar Energy and Hydrogen Fuel Cells
> 3: Wind Energy, Oceanic Energy, and Hydropower
> 4: Geothermal and Biomass Energy
> 5: Energy Efficiency, Conservation, and Sustainability

Mining: of coal, **1:**106–12; environmental issues of, **1:**109–11; of oil shale, **1:**58; surface, **1:**107–9; underground, **1:**108–9; uranium, **1:**138–40

Minnesota: carbon footprint reduction in, **5:**9–10; geothermal systems used in, **4:**66; wind farms in, **3:**34

Minnesota Schools Cutting Carbon project, **5:**9, **5:**10*f*

Miravalles volcano power station, **4:**43

Mississippi, **4:**72

Missouri, **3:**29, **3:**36

Mitsubishi, **2:**102, **5:**71

Mochida, Hiroko, **5:**118*f*

Model cars, solar energy, **1:**87*f*, **1:**88

Model racing cars, **2:**101–2

Model T hot rod, **2:**95–99, **2:**95*f*

Moderators, **1:**141–42

Modernization, **5:**27–28

Moeller, Keats, **1:**64–65

Mojave Desert, **2:**36–38, **2:**51, **3:**32, **4:**11

Molecules, **2:**10

Molina, Raquel, **1:**50

Moller, Kris, **4:**109, **4:**127

Molten Carbonate fuel cells (MCFC), **2:**119

Molten salt storage, **2:**39, **2:**48–49

Monarch School, **5:**11

Mongillo, John, **5:**125–26, **5:**126*f*

Montana, **4:**20

Moos Lake water-treatment plant, **5:**83

Motion, **1:**6

Mouchout, Auguste, **1:**198, **1:**206, **2:**168, **2:**176, **3:**166, **3:**174, **4:**168, **4:**176, **5:**170, **5:**178

Mount Washington Cog Railway, **4:**108

Mt. Washington, **3:**5

Muddy Run Pumped Storage Facility, **3:**81

Murphy, John, **2:**98

Museum of Science, **3:**26

Musicians, **5:**6–8, **5:**8*f*

Musk, Elon, **1:**206, **2:**176, **3:**174, **4:**176, **5:**178

Nacelle, **3:**11

Nanocoatings Subscale Laboratory, **5:**117*f*

Nanometers, **2:**10

Nanosolar, **1:**215, **2:**185, **3:**183, **4:**185, **5:**187

Nano Solar Technology, **2:**56

Nanotechnology, **4:**77, **5:**114–15, **5:**114*f*, **5:**116

Nanotechnology and Energy, **5:**128

Naruse, Masanori, **2:**118

NASA. *See* National Aeronautics and Space Administration

NASCAR, **5:**1; France, Bill, of, **5:**2*f*; go-green program of, **5:**1–3

National Aeronautics and Space Administration (NASA), **1:**16*f*, **2:**90, **5:**120

National Association for Stock Car Auto Racing. *See* NASCAR

National Biodiesel Board (NBB), **4:**117, **4:**127

National Earth Comfort Program, **4:**73

National Energy Education Development (NEED), **1:**154, **2:**52, **3:**64–68; mission of, **3:**66; real world issues addressed by, **3:**66–67

National Energy Foundation, **1:**68

National Energy Technology Laboratory (NETL), **1**:123, **1**:125
National Fuel Cell Research Center, **2**:103
National Gas Supply Association (NGSA), **1**:100
National Geographic Society's Green Guide, **5**:90
National Hydropower Association, **1**:36, **3**:100
National Oceanic and Atmospheric Administration (NOAA), **3**:124
National parks, **4**:52
National Renewable Energy Laboratory (NREL), **1**:166, **2**:9, **2**:30, **2**:33, **2**:99–100, **2**:103, **2**:127, **2**:136, **3**:14, **3**:44, **3**:134, **4**:118, **4**:136, **5**:109, **5**:115, **5**:138
National science education standards, **1**:217–18, **2**:187–88, **3**:185–86, **4**:187–88, **5**:189–90
National Solar Bike Rayce, **2**:45
National Wind Technology Center (NWTC), **3**:52
Native American tribes, **2**:8
Natural Energy Laboratory of Hawaii Authority, **3**:119
Natural environment, **5**:80
Natural gas, **1**:xii, **1**:9–10, **1**:71–72, **2**:xii, **3**:xii, **4**:xii, **5**:xii, **5**:124; Alaska North Slope's deposits of, **1**:83; benefits of, **1**:93; China extracting, **1**:71; China's use of, **1**:197, **2**:167, **3**:165, **4**:167, **5**:169; as cleanest fossil fuel, **1**:94; in Columbia, **1**:101; consumers of, **1**:82–85; consumption, **1**:83–84; contents of, **1**:74–75; drilling for, **1**:76–77; emission levels and, **1**:90–94; formation of, **1**:74; France and, **1**:84; future of, **1**:94–95; Germany's consumption of, **1**:83–84; global consumption of, **1**:72–73, **1**:94–95; global reserves of, **1**:79–81, **1**:79*f*, **1**:188*t*–189*t*, **2**:158*t*–159*t*, **3**:156*t*–157*t*, **4**:158*t*–159*t*, **5**:160*t*–161*t*; greenhouse gases and, **1**:94; history of, **1**:73–74; industries use of, **1**:72; industry of, **1**:78*f*; industry using, **1**:72; locating deposits of, **1**:76; measurement of, **1**:78–79; in Middle East, **1**:79*f*; Netherlands and, **1**:84; new drilling technologies for, **1**:77; Norway's reserves of, **1**:82; pipeline transportation of, **1**:77–78; power plant for, **1**:73*f*; production, **1**:81–82; Russia's reserves of, **1**:82; steam reformation from, **2**:91; synthetic, **1**:125–28; United Kingdom's consumption of, **1**:84; US consumption of, **1**:83
Natural Gas Star Program, **1**:94
Natural gas vehicles (NGVs), **1**:85–86, **1**:89–92, **5**:75; advantages and disadvantages of, **1**:91–92; safety of, **1**:92
Natural lighting, **5**:37, **5**:105
Natural resources, **1**:25–26
Natural Resources Research Institute (NRRI), **4**:83
Nauen, Andreas, **1**:207, **2**:177, **3**:175, **4**:177, **5**:179
Naval Petroleum and Oil Shale Reserves, **1**:58
NBB. *See* National Biodiesel Board
Nebraska, **4**:68
NECAR 1, **2**:90
NEED. *See* National Energy Education Development
NEI. *See* Nuclear Energy Institute
Nellis Air Force Base, **2**:3–4, **2**:3*f*, **2**:13
NESEA. *See* Northeast Sustainable Energy Association
Netherlands, **1**:84
NETL. *See* National Energy Technology Laboratory
Net metering, **2**:15–16, **3**:43, **3**:71
NEUP. *See* Nuclear Energy University Program

> 1: Oil, Natural Gas, Coal, and Nuclear
> 2: Solar Energy and Hydrogen Fuel Cells
> 3: Wind Energy, Oceanic Energy, and Hydropower
> 4: Geothermal and Biomass Energy
> 5: Energy Efficiency, Conservation, and Sustainability

"Neutropolis: The Nuclear Energy Zone for Students," **1**:152
Nevada, **4**:15
Nevada Solar One, **1**:215, **2**:38–39, **2**:185, **3**:183, **4**:185, **5**:187
Newell, Craig, **3**:20
New Hampshire, **5**:10–11
New Jersey, **4**:118, **5**:35–36
New Mexico, **4**:17–18
New Planet Energy, **5**:128
Newsom, Gavin, **2**:112
New York, **2**:17, **2**:114–15, **5**:34, **5**:82
New York gym, **5**:9
New Zealand, **1**:85; geothermal power station in, **4**:44*f*; geothermal resources in, **4**:44; tidal power in, **3**:110–11
NGSA. *See* National Gas Supply Association
NGVs. *See* Natural gas vehicles
Niagara Falls, **3**:76
Nicholson, William, **2**:90
Nickel-metal hydride battery (NiMH), **5**:122–23
Nielsen, Carl, **4**:6
NiMH. *See* Nickel-metal hydride battery
NIRS. *See* Nuclear Information and Resource Service
Nissan Leaf electric car, **5**:74
Nissan Motor Co., **5**:71, **5**:72
Nitrogen oxide, **1**:98
NOAA. *See* National Oceanic and Atmospheric Administration

Nonrenewable energy, **1**:xi–xiii, **1**:9–11, **2**:xi–xiii, **3**:xi–xiii, **4**:xi–xiii, **5**:xi–xiii; career resources in, **1**:209–12, **2**:179–82, **3**:177–80, **4**:179–82, **5**:181–84
Non-silicon-based technologies, **2**:11
Norman, Marie, **5**:28–31, **5**:29*f*
Northbrook High School, **5**:57–62
North Dakota, **4**:72
Northeast Blackout of 1965, **1**:1, **1**:2*f*
Northeast Sustainable Energy Association (NESEA), **1**:88
Northeast US, **3**:36–37
Northern Ireland, **3**:110
North Grand High School, **4**:122–25
North Sea, **1**:80*f*
Norway: hydroelectric energy in, **1**:xvi, **2**:xvi, **3**:xvi, **3**:84*f*, **3**:87, **4**:xvi, **5**:xvi, **5**:112; hydrogen refueling in, **2**:114, **2**:115*f*; natural gas reserves of, **1**:82; wave energy used in, **3**:117–18
NRC. *See* Nuclear Regulatory Commission
NREL. *See* National Renewable Energy Laboratory
NRG Energy, Inc, **2**:22
NRRI. *See* Natural Resources Research Institute
Nuclear energy, **1**:xiii, **1**:5, **1**:10–11, **2**:xiii, **3**:xiii, **4**:xiii, **5**:xiii; benefits of, **1**:132–33; in countries, **5**:112; description of, **1**:133–34; in France, **1**:137–38; fuel rods in, **1**:140–41, **1**:141*f*; functioning of, **1**:158; future of, **1**:152–53; greenhouse gases and, **1**:11; history of, **1**:134; nuclear fission in, **1**:139; reading materials on, **1**:159–60, **2**:129–30, **3**:127–28, **4**:129–30, **5**:131–32; in sustainable development, **5**:110–12, **5**:111*f*; technologies in, **1**:151–52; in US, **1**:132–33, **1**:137, **5**:112; US military use of, **1**:137; world's electrical

needs and, **1:**11; world use of, **1:**136–38; as zero-carbon energy source, **1:**131–32

Nuclear Energy Institute (NEI), **1:**152, **1:**157

Nuclear Energy University Program (NEUP), **1:**131, **5:**111

Nuclear fission, **1:**134, **1:**139

Nuclear fuel: cycle, **1:**132; nuclear waste transformed to, **1:**150–51; uranium mining and, **1:**138–40

Nuclear Information and Resource Service (NIRS), **1:**157

Nuclear power plants, **1:**200, **2:**170, **3:**168, **4:**170, **5:**172; Browns Ferry, **1:**201, **2:**171, **3:**169, **4:**171, **5:**173; floating, **1:**151–52; Sizewell, **1:**133; waste generated by, **1:**145–51

Nuclear reactors: in China, **1:**11*f*; designing, **1:**153–54; manufacturers of, **1:**145; types of, **1:**142–44; in US, **1:**136

Nuclear Regulatory Commission (NRC), **1:**134, **1:**147, **1:**157, **1:**166, **2:**136, **3:**134, **4:**136, **5:**138

Nuclear waste: disposing of, **1:**147–48; high-level, **1:**145; low-level, **1:**145–46; nuclear fuel transformed from, **1:**150–51; of power plants, **1:**145–51; recycling of, **1:**149, **1:**151; transuranic, **1:**146–47

Nuclear Waste Policy Act, **1:**148–49

Nuclear weapons, **1:**135

NWTC. *See* National Wind Technology Center

Oak Ridge National Laboratory, **1:**150, **5:**123

Obama, Barack, **1:**xiii, **1:**51, **1:**123, **1:**131, **1:**148, **2:**3, **2:**3*f*, **2:**86, **3:**xiii, **4:**xiii, **4:**80, **5:**xiii, **5:**108*f*, **5:**110–11, **5:**124, **5:**129

Ocean Energy Council, **3:**124

Ocean Power Technologies, **3:**114

Ocean Renewable Power Company (ORPC), **3:**124

Oceans: kinetic energy from, **3:**103–4; as solar energy collector, **3:**108; thermal energy from, **3:**118–23; tidal technologies and, **3:**124–25; wave energy from, **3:**113–18

Ocean Thermal Energy Conversion (OTEC), **1:**15, **3:**104, **3:**118–19, **3:**120*f*; Africa and, **3:**122; challenges facing, **3:**122; defining, **3:**119; future of, **3:**122–23; technologies in, **3:**119–22

Octane rating, **1:**47

OECD. *See* Organization of Economic Cooperation and Development

Oerlikon Solar, **1:**215, **2:**185, **3:**183, **4:**185, **5:**187

Office of Energy Efficiency and Renewable Energy, **4:**73

Office of Fossil Energy, **1:**77

Offshore drilling, **1:**80*f*

Offshore wave energy generation systems, **3:**114

Ohio, **5:**33

Ohms Law, **1:**198, **2:**168, **3:**166, **4:**168, **5:**170

Oil: algae high in, **2:**93; carbon ratio of, **1:**94; deposits, **1:**43; discarded restaurant, **4:**116*f*; drilling rig, **1:**44*f*; enhanced recovery of, **1:**45; extraction of, **1:**63; fields, **1:**43, **1:**55; gas production and, **1:**62; industry, **3:**16; peak, **1:**56, **1:**69; –producing countries, **1:**56; –producing states, **1:**55*f*; recovery of, **1:**45; refineries, **1:**20; reserves remaining of, **1:**62–63; spills, **1:**61; thermal recovery of, **1:**45; US fields of, **1:**46; US imports of, **1:**201, **2:**171, **3:**169, **4:**171,

1: Oil, Natural Gas, Coal, and Nuclear
2: Solar Energy and Hydrogen Fuel Cells
3: Wind Energy, Oceanic Energy, and Hydropower
4: Geothermal and Biomass Energy
5: Energy Efficiency, Conservation, and Sustainability

5:173; world, **1**:190, **2**:160, **3**:158, **4**:160, **5**:162
Oil and Gas Journal, **1**:66
Oil Pollution Act, **1**:62
Oil sands, **1**:58–59, **1**:190, **2**:160, **3**:158, **4**:160, **5**:162
Oil shale, **1**:57–58; mining of, **1**:58; US deposits of, **1**:57*f*
Old Faithful, **4**:4, **4**:4*f*
Olive oil, **1**:197, **2**:167, **3**:165, **4**:167, **5**:169
Olmedilla Photovoltaic Park, **2**:24
Online Fuel Cell Information Resource, **2**:127
Onshore wave energy systems, **3**:114–16
On-site fuel cells, **5**:119
OPEC. *See* Organization of Petroleum Exporting Countries
Open-cycle systems, **3**:121
Open-loop systems: disadvantages of, **4**:61–62; geothermal heat pump as, **4**:61–63, **4**:62*f*
ORC. *See* Organic Rankine cycle
Oregon Institute of Technology, **4**:1–2, **4**:20–23, **4**:21*f*, **4**:24*f*
Organic matter, **1**:74
Organic Rankine cycle (ORC), **4**:21*f*, **4**:24*f*
Organic vegetable waste, **4**:86
Organization of Economic Cooperation and Development (OECD), **1**:20*f*, **1**:81

Organization of Petroleum Exporting Countries (OPEC), **1**:56, **1**:166, **2**:136, **3**:134, **4**:136, **5**:138
Organizations, government websites and, **1**:165–67, **2**:135–37, **3**:133–35, **4**:135–37, **5**:137–39
Ormat, **4**:20
ORPC. *See* Ocean Renewable Power Company
Oscillating water column converter, **3**:115
OTEC. *See* Ocean Thermal Energy Conversion
Overmann, Harold, **3**:20
Oxford Yasa Motors, **1**:215, **2**:185, **3**:183, **4**:185, **5**:187
Oxygenates, **1**:53

Pacific Fuel Cell Company, **2**:102
Pacific Gas and Electric, **3**:67
Pacific ocean, **4**:6–7
PAFCs. *See* Phosphoric acid fuel cells
Paint Lick Elementary School, **4**:65
Palapa, Rosy, **1**:50
Panjshir River, **3**:95*f*
Paper manufacturing, **1**:21
Parabolic cookers, **2**:72
Parabolic solar oven, **2**:71
Parabolic trough system, **2**:36–39, **2**:37*f*, **2**:43*f*
Paraguay, **3**:86–87
Parr, Alexandria, **1**:30
Passamaquoddy Bay, **3**:109
Passive heating and cooling, **1**:26
Passive Solar Design, **2**:83
Passive solar energy, **2**:62
Passive solar heating systems, **2**:67–70, **2**:67*f*; benefits of, **2**:70; home design with, **2**:68–69; school design with, **2**:69–70
Passive solar technology, **2**:60–61
Passive solar water heaters, **2**:77*f*
Passive yawing, **3**:12

Paul, Stephen, **1:**207, **2:**177, **3:**175, **4:**177, **5:**179
PBMR. *See* Pebble bed modular reactor
Peak oil, **1:**56, **1:**69
Pearl Street Station, **1:**199, **2:**169, **3:**167, **4:**169, **5:**171
Pearson, Gerald, **2:**5
Pebble bed modular reactor (PBMR), **1:** 151
Peck, Rick, **5:**13–15, **5:**13*f*
Pelamis Wave Power, Ltd., **3:**114
Pelton, Lester, **3:**93*f*
Pelton and Turgo impulse turbines, **3:**92–93, **3:**93*f,* **3:**94
PEM. *See* Polymer electrolyte membrane
Pendulor devices, **3:**115
Penn State Green Roof Research Center, **5:**96
Pennsylvania, **4:**121, **5:**31
Penstock, **3:**90, **3:**92*f*
Petrochemicals, **1:**53
Petroleum, **1:**9; biodiesel emissions compared to, **4:**116; defining, **1:**43; drilling for, **1:**43–45; early uses of, **1:**197, **2:**167, **3:**165, **4:**167, **5:**169; environmental issues concerning, **1:**60–62; forming of, **1:**41–42; as fossil fuel, **1:**41; future of, **1:**63; history of, **1:**40–41; locating, **1:**68; petrochemicals from, **1:**53; present uses of, **1:**39–40; products from, **1:**47–48; recovery of, **1:**45; searching for, **1:**42–43; US consuming, **1:**60; US importing, **1:**41; world production of, **1:**xi–xii, **2:**xi–xii, **3:**xi–xii, **4:**xi–xii, **5:**xi–xii
Philippines, **4:**32–34, **4:**87
Phone chargers, **5:**54
Phosphoric acid fuel cells (PAFCs), **2:**119
Photosynthesis, **2:**11, **2:**92–93

Photovoltaic cells (PV), **1:**xiii, **1:**12, **2:**xiii, **2:**1, **3:**xiii, **4:**xiii, **5:**xiii; to battery storage, **2:**14; disposal and recycling of, **2:**30; installation of, **2:**19–20, **2:**22*f;* materials used in, **2:**6; sized and shapes of, **2:**12–13; uses for, **2:**6
Photovoltaic power plants, **2:**16*f*
Photovoltaic system, **5:**34, **5:**34*f,* **5:**113
Photovoltaic technology: Australia using, **2:**25; future of, **2:**30–31; Germany using, **2:**24–25; global leaders in, **2:**24–26
Physical science, **1:**217, **2:**187, **3:**185, **4:**187, **5:**189
Pickens, T. Boone, **3:**27
Picohydro, **3:**91
Pipeline transportation, **1:**77–78
Pittsburgh National Corporation, **5:**88
Plate-boundary volcanoes, **4:**7*f*
Pleasant Ridge Montessori School, **5:**33
Plutonium uranium recovery by extraction (PUREX), **1:**150–51
Poland, **1:**116
Polymer electrolyte membrane (PEM), **1:**xv, **2:**xv, **2:**89, **3:**xv, **4:**xv, **5:**xv
Pools, **4:**50–51
Portland, Maine, **3:**105*f,* **5:**90–91
Portsmouth Abbey monastery, **3:**37
Portugal: wave energy in, **3:**117; wind energy in, **3:**59
Potential energy, **1:**5, **3:**8
Powder River Basin, **1:**114
PowerBuoy, **3:**114
Power plants: Beaver County, **4:**20; binary, **4:**9–10, **4:**9*f,* **4:**22; CO_2 producing, **1:**73*f;* diversion, **3:**81; double-flash, **4:**11; flash-steam, **4:**10–12, **4:**10*f;* floating

> 1: Oil, Natural Gas, Coal, and Nuclear
> 2: Solar Energy and Hydrogen Fuel Cells
> 3: Wind Energy, Oceanic Energy, and Hydropower
> 4: Geothermal and Biomass Energy
> 5: Energy Efficiency, Conservation, and Sustainability

nuclear, **1:**151–52; hydrogen, **2:**94; La Rance River tidal, **3:**109*f;* Mammoth Pacific, **4:**10; for natural gas, **1:**73*f;* nuclear waste of, **1:**145–51; offshore wind energy, **3:**62–64; pumped storage, **3:**81; Raft River, **4:**15–16, **4:**16*f;* Raser Technologies thermo, **4:**19*f;* Velling Mærsk-Tændpibe wind, **3:**59; Voith Siemens Hydro, **5:**108*f. See also* Geothermal power plants; Hydroelectric power plants; Microhydroelectric power plants

Power strips, **5:**54
Power Technology, **2:**83
Pressures, **4:**3
Pressurized water reactors, **1:**143, **1:**144
Primary footprint, **5:**6
Princeton Review, **5:**42
Prism Solar Technologies, **2:**33
Probst, Pete, **4:**124
Product development, in energy, **1:**213–16, **2:**183–86, **3:**181–84, **4:**183–86, **5:**185–88
Production: biodiesel specifications of, **4:**114–16; biogas facilities for, **4:**97*f;* China's coal, **1:**114, **1:**116; coal, **1:**115–16, **1:**116; countries coal, **1:**115–16; ethanol, **4:**93–95; fuel cells, **2:**104; gas and oil, **1:**62; hydrogen, **2:**91–100; India's coal, **1:**116; natural gas, **1:**81–82; Poland's coal, **1:**116; renewable energy consumption and, **1:**180*t*–183*t,* **2:**150*t*–153*t,* **3:**148*t*–151*t,* **4:**150*t*–153*t,* **5:**152*t*–155*t;* residential systems, **3:**16, **3:**41; thin-film solar cells electricity, **2:**9–10; wind energy, **3:**29–38, **3:**56–58, **3:**58*f,* **5:**115; wind turbines energy, **3:**42

Products: coal made in to, **1:**105*f;* from crude oil, **1:**40*f;* from petroleum, **1:**47–48; refined, **1:**59
Project Driveway, **2:**114
Project Two Degrees, **5:**18
The Promise of Solar Energy, **2:**83
Propane, **1:**74–75, **1:**76, **1:**85
Puertollano Photovoltaic Park, **2:**24
Pulverized coal-fired burners, **1:**120
Pumped storage power plant, **3:**81
Puna Geothermal Venture, **4:**16–17
PUREX. *See* Plutonium uranium recovery by extraction
PV. *See* Photovoltaic cells
PV Crystalox Solar, **1:**216, **2:**186, **3:**184, **4:**186, **5:**188

Quad, **1:**18
Quantum dots, **5:**114–15
Quantum Technologies, **2:**122
Quebec, **3:**86

Race cars, **4:**107–8, **4:**108*f*
Radial drilling, **4:**59, **4:**77
Radiant energy, **1:**5
Radiation, **2:**61, **2:**68
Raft River power plant, **4:**15–16, **4:**16*f*
Rain barrels, **5:**32
Ramirez, Daniel, **1:**50
Rance estuary, **3:**106
Ranking criteria, **5:**90
Raser Technologies, **4:**19*f,* **4:**20
Rawal, Bhavna, **5:**57–62, **5:**58*f*

Reading materials: on biomass, **1:**162, **2:**132, **3:**130, **4:**132, **5:**134; on energy conservation, **1:**162–63, **2:**132–33, **3:**130–31, **4:**132–33, **5:**134–35; on energy efficiency, **1:**162–63, **2:**132–33, **3:**130–31, **4:**132–33, **5:**134–35; on fossil fuels, **1:**159–60, **2:**129–30, **3:**127–28, **4:**129–30, **5:**131–32; on geothermal energy, **1:**162, **2:**132, **3:**130, **4:**132, **5:**134; on hydroelectric energy, **1:**161, **2:**131, **3:**129, **4:**131, **5:**133; on hydrogen fuel cells, **1:**160–61, **2:**130–31, **3:**128–29, **4:**130–31, **5:**132–33; on nuclear energy, **1:**159–60, **2:**129–30, **3:**127–28, **4:**129–30, **5:**131–32; on solar energy, **1:**160–61, **2:**130–31, **3:**128–29, **4:**130–31, **5:**132–33; on sustainable development, **1:**162–63, **2:**132–33, **3:**130–31, **4:**132–33, **5:**134–35; on wave energy, **1:**161, **2:**131, **3:**129, **4:**131, **5:**133; on wind energy, **1:**161, **2:**131, **3:**129, **4:**131, **5:**133

Reclamation, surface mining and, **1:**107–8

Recoverable reserves, of coal, **1:**190–91*t*, **2:**160*t*–161*t*, **3:**158*t*–159*t*, **4:**160*t*–161*t*, **5:**162*t*–163*t*

Recycling, **1:**25; of materials, **5:**7, **5:**104, **5:**109; of nuclear waste, **1:**149, **1:**151; of PV, **2:**30

REEEP. *See* Renewable Energy and Energy Efficiency Partnership

Refined products, **1:**59

Refineries, crude oil, **1:**46–47

Refrigerant, liquid, **4:**59

Refueling of FCVs, **2:**114–15

Regional energy consumption, **1:**184*t*–187*t*, **2:**154*t*–157*t*, **3:**152*t*–155*t*, **4:**154*t*–157*t*, **5:**156*t*–159*t*

Regulator, **3:**90

Reid, Harry, **1:**148, **2:**3*f*

Renewable energy, **1:**12–16, **1:**14*f*, **5:**23, **5:**30; career resources in, **1:**209–12, **2:**179–82, **3:**177–80, **4:**179–82, **5:**181–84; China promoting, **3:**61*f*; Database of State Incentives for, **3:**71; future of, **5:**124; for homes, **5:**63–64; interstate coalition for, **2:**16; Native American tribes providing, **2:**8; primary energy sources and, **1:**180*t*–183*t*, **2:**150*t*–153*t*, **3:**148*t*–151*t*, **4:**150*t*–153*t*, **5:**152*t*–155*t*; production and consumption of, **1:**180*t*–183*t*, **2:**150*t*–153*t*, **3:**148*t*–151*t*, **4:**150*t*–153*t*, **5:**152*t*–155*t*; resources of, **1:**xiii–xviii, **2:**xiii–xviii, **3:**xiii–xviii, **4:**xiii–xviii, **5:**xiii–xviii; seaweed as, **1:**98; sustainable future powered by, **5:**107–22; transmission of, **2:**28–29

Renewable Energy Act, **4:**87

Renewable Energy and Energy Efficiency Partnership (REEEP), **4:**34

Renewable Resource Data Center (RReDC), **5:**125

Renovation, **5:**27–28

Residential systems: annual production of, **3:**16, **3:**41; fuel cell, **5:**118–19; grid connection of, **3:**42–43, **3:**42*f*; net metering with, **3:**43; small wind turbines for, **3:**40–42; using wind energy, **3:**40–44; wind turbine maintenance of, **3:**41

Revenues, **3:**38

Reverb, **5:**8*f*

Reykjavik, Iceland, **4:**25

Rhode Island, **3:**37, **4:**118–19, **5:**126

Rhode Island Resource Recovery, **2:**98

Rice paddies, **1:**76

> 1: Oil, Natural Gas, Coal, and Nuclear
> 2: Solar Energy and Hydrogen Fuel Cells
> 3: Wind Energy, Oceanic Energy, and Hydropower
> 4: Geothermal and Biomass Energy
> 5: Energy Efficiency, Conservation, and Sustainability

Richardson, Bill, **4:**18
Richmond BUILD, **1:**30
Ring of Fire, **4:**6–7, **4:**7*f*, **4:**18, **4:**31, **4:**32, **4:**33; active volcanoes in, **4:**40, **4:**44
Rising Sun Energy Center, **1:**29, **1:**31
River control projects, **3:**79
Riversimple hydrogen car, **2:**110, **2:**110*f*, **5:**69
Robbins, Steve, **2:**9
Rocket stoves, **4:**88
Rock Port, Missouri, **3:**29
Rodriguez, Jose, **1:**50
Rojas, Fermin, **1:**50
Romania, **3:**88–89
Roof is Growing, **5:**95
Roofscapes, Inc., **5:**96
Rooftop gardens, **5:**78, **5:**84, **5:**87
Rotary drilling rigs, **1:**43, **1:**76–77
Rotor blades, **3:**7–9
Roy Lee Walker Elementary School, **5:**31
RReDC. *See* Renewable Resource Data Center
Rudensey, Lyle, **4:**116*f*
Russia: natural gas reserves of, **1:**82; wind energy potential of, **3:**62

Safety-Kleen, Inc., **5:**2
SAI. *See* Solar America Initiative
Saint Thomas Academy, **2:**44–45, **2:**44*f*
SAITEM. *See* Sakarya University Advanced Technologies Implementation Group
Sakarya University Advanced Technologies Implementation Group (SAITEM), **2:**116
Salter, Steven, **1:**207, **2:**177, **3:**175, **4:**177, **5:**179
Sandia National Laboratories, **1:**216, **2:**186, **3:**9, **3:**184, **4:**29, **4:**186, **5:**28, **5:**188
San Diego High School, **4:**121
San Domenico School, **5:**33
Sandstone, **1:**42
San Francisco, **3:**111, **4:**5–6
San Francisco International Airport, **2:**111
Sangre de Cristo Mountains, **5:**47*f*
San Jose, California, **5:**90
San Juan Basin, **1:**80
Santa Coloma de Gramenet cemetery, **2:**25
Santa Monica, California, **5:**17
Saturn, **5:**71
Saturn EV-1 electric car, **5:**72, **5:**122
Savannah River National Laboratory, **2:**94
Schaefer, Natasha, **5:**27*f*
School building: green roof of, **5:**38–40, **5:**38*f*; renovation and modernization of, **5:**27–28; sustainable green, **5:**26
School Planning & Management, **2:**69
Schools: biodiesel buses for, **4:**117–20, **4:**119, **4:**119*f*; biodiesel vehicles used by, **4:**117–20; carbon footprint reduction of, **5:**12, **5:**14–15; carbon footprints of, **5:**9–12; eco-friendly, **4:**65*f*; energy conservation and efficiency in, **1:**26–28; energy efficiency in, **1:**26–28; energy efficient programs for, **5:**45; Fort Atkinson School District, **4:**66–67; fuel cell education in, **2:**125–26; go-green

public, **5:**31–36; green certification of, **5:**27*f*; heat pump benefits to, **4:**64; Indiana's geothermal well used in, **4:**68; LEED certification for, **5:**28–32, **5:**82; Massachusetts' geothermal heat pump used in, **4:**67–68; Minnesota's geothermal systems in, **4:**66; natural lighting in, **5:**37; Nebraska and geothermal heat pump in, **4:**68; passive solar design for, **2:**69–70; South Dakota's geothermal energy in, **4:**67; using geothermal heat pump, **4:**64–68; using solar energy, **2:**16–17; utility bills lowered in, **5:**36; wind energy in, **3:**35*t*, **3:**44, **3:**46; wind turbines for, **3:**16–23; Wisconsin's geothermal systems in, **4:**66
SchoolsCuttingCarbon.org, **5:**10
Schools for Energy Efficiency, **5:**23
Schwarzenegger, Arnold, **3:**32
Schwarze Pumpe, **1:**124*f*
Science, **1:**218, **2:**188, **3:**186, **4:**188, **5:**190
Science education, **1:**217, **2:**64, **2:**187, **3:**185, **4:**187, **5:**189
SciKits, **2:**127
Scooters, **2:**113–14
Scott, Allister, **4:**123
Scott, Paul, **5:**18*f*
SEAT. *See* Student Energy Audit Training
Seattle, Washington, **5:**89*f*
Seaweed, **1:**98
Secondary footprint, **5:**6
Sector, energy consumption by, **1:**174*t*–177*t*, **2:**144*t*–147*t*, **3:**142*t*–145*t*, **4:**144*t*–147*t*, **5:**146*t*–149*t*
Sedum, **5:**39
SEGS. *See* Solar Energy Generating Systems
SEI. *See* Solar Energy International

Selsam, Douglas, **1:**207, **2:**177, **3:**175, **4:**177, **5:**179
Seneca Ridge Middle School, **5:**13–15, **5:**13*f*
Sequestration technology, **1:**125
Sharp Corp., **2:**26
Shell Hydrogen Fuel, **2:**114
Shell WindEnergy, Inc., **3:**31
Shippingport Atomic Power Station, **1:**135
Shutol microhydroelectric power plants, **3:**95*f*
Siemens Corp., **1:**216, **2:**186, **3:**184, **4:**186, **5:**188
Silicon, **2:**9
Simón Bolivar hydroelectric power plants, **3:**87
Single-flash geothermal power plants, **4:**11
Site evaluation, **4:**63–64
Sizewell nuclear power plant, **1:**133
Skystream 3.7 wind turbine, **3:**46*f*
Small Business Guide to Energy Efficiency, **5:**23
Smaller homes, **5:**55
Small-scale hydroelectric power plants, **3:**89–90
SmartWay designation, **5:**75
Smith, David, **5:**48
Smith, Wylie, **4:**110*f*, **4:**111
Smith Family Dome Homes, **5:**48
Snodgrass, Edmund C., **5:**40
Snodgrass, Lucie L., **5:**40
SOFC. *See* Solid oxide fuel cell
Solar America Cities, **2:**21
Solar America Initiative (SAI), **2:**31
Solar arrays: at Buckley Air Force Base, **2:**14; at Hanneford Supermarket, **5:**103*f*; power of, **2:**13; SunPower, **2:**18*f*
Solar cells: China's manufacturing of, **1:**xiv, **1:**201, **2:**xiv, **2:**24, **2:**171, **3:**xiv, **3:**169, **4:**xiv, **4:**171, **5:**xiv, **5:**173; crystalline silicon, **2:**7–8;

1: Oil, Natural Gas, Coal, and Nuclear
2: Solar Energy and Hydrogen Fuel Cells
3: Wind Energy, Oceanic Energy, and Hydropower
4: Geothermal and Biomass Energy
5: Energy Efficiency, Conservation, and Sustainability

dye-sensitized technology of, **2:**11–12, **2:**12*f*; electricity produced by, **2:**7*f*; from fruit, **2:**13; functioning of, **2:**7–8; nanotechnology used in, **5:**114*f*; new generation of, **2:**8–10; non-silicon-based technologies in, **2:**11; quantum dots with, **5:**114–15; thin-film, **2:**9–10, **2:**9*f*
Solar Decathlon, 2009 homes in, **5:**64–65, **5:**64*f*
Solar dish-engine system, **2:**39–42
Solar energy, **1:**xiii–xv, **1:**12, **2:**xiii–xv, **3:**xiii–xv, **4:**xiii–xv, **5:**xiii–xv; Abengoa Solar and, **1:**213, **2:**40, **2:**55, **2:**183, **3:**181, **4:**183, **5:**185; Africa using, **2:**26; availability of, **2:**4–5; benefits of, **2:**26–27; California using, **2:**17; cities using, **2:**17–21; collectors for, **2:**73–74, **2:**74*f*; college installation of, **2:**4; cost of, **2:**30–31; economics of, **2:**30; electricity created by, **2:**14–15; future, **2:**83; history of, **2:**5–6, **2:**56; for homes, **5:**63–64; Hong Kong ferry boats using, **2:**6; Japan and, **2:**83; Japan's cargo ships using, **2:**27; Japan's installations of, **2:**25–26; land use and, **2:**29; limiting factors of, **2:**27–29; model cars using, **1:**87*f*, **1:**88; net metering and, **2:**15–16; oceans collector of, **3:**108; passive, **2:**62; photovoltaic power plants for, **2:**16*f*; reading materials on, **1:**160–61, **2:**130–31, **3:**128–29, **4:**130–31, **5:**132–33; schools using, **2:**16–17; science of, **2:**61–62; solar tower plant and, **2:**48*f*; Southwestern farms for, **2:**23; Spain's investments in, **2:**24; storage of, **2:**13–14, **2:**28; in sustainable development, **5:**112–15; transmission gridlines for, **2:**28–29; US installations of, **2:**3–4; US projects of, **2:**21–23; US using, **2:**17–21; utility grid and, **2:**14–15, **2:**28; videos for, **2:**34. *See also* Photovoltaic cells; Photovoltaic technology
Solar Energy Generating Systems (SEGS), **2:**37, **2:**51, **2:**51*f*
Solar Energy International (SEI), **2:**8
Solar Energy Review, **2:**82
Solar greenhouse, **2:**65–66
Solar heaters, **2:**79*f*
Solar heating systems, **2:**66–67
Solar oven: benefits of, **2:**71–72; parabolic, **2:**71; Thames and Cosmos, **2:**73*f*
Solar ovens, **2:**71
Solar panels: A.A. Kingston Middle School with, **5:**34*f*; in Germany, **2:**27*f*; of Hanneford Supermarket, **5:**103–4; Spain's installation of, **2:**25; temperature influencing, **3:**64*f*
Solar ponds, **2:**80–82
Solar powered vehicles, **2:**45–46
Solar power plant, **2:**39
Solar power tower system, **2:**42–47
Solar radiation, **2:**5, **2:**61–62
Solar schools program, **3:**67
Solar thermal electric (STE), **2:**38*fs*
Solar towers, **2:**56, **2:**83; benefits of, **2:**49–50; molten salt storage with, **2:**48–49; solar energy from, **2:**48*f*; of Spain, **2:**47
Solar Two, **2:**47–50
Solar wall, **2:**60–61

Solar water heaters, **1**:xiv, **2**:xiv, **2**:24, **3**:xiv, **4**:xiv, **5**:xiv, **5**:66–68; active, **2**:76*f;* cost and benefits of, **2**:79, **5**:66–67; economics of, **2**:78; evolution of, **2**:75; global installations of, **2**:78–79; greenhouse gas emissions reduced by, **2**:78; for homes, **5**:64; passive, **2**:77*f;* storage tanks required by, **2**:77; students designing, **2**:80; types of, **2**:75–77; world's first, **2**:59
Solatubes, **5**:37
Solid oxide fuel cell (SOFC), **2**:107, **5**:121
Solix Biofuels, **1**:216, **2**:186, **3**:184, **4**:186, **5**:188
Solvents, **1**:47
Sony Corporation, **2**:12*f*
Soultz-sous-Forêts, **4**:48, **4**:48*f*
Sound, **1**:6
Source, energy consumption by, **1**:170*t*–173*t*, **2**:140*t*–143*t*, **3**:138*t*–141*t*, **4**:140*t*–143*t*, **5**:142*t*–145*t*
South America, **1**:84
South Dakota: geothermal energy in, **4**:67; wind farms in, **3**:36
Southeast Asia, **3**:61
South Korea, **3**:110
Southwestern solar farms, **2**:23
Soybean-powered buses, **4**:109*f*
Soybean-powered cars, **4**:121
Soy crops, **4**:125
Space Conditioning: The Next Frontier, **4**:69
Space science, **1**:218, **2**:188, **3**:186, **4**:188, **5**:190
Space shuttles, **5**:120
Spain: solar energy investments of, **2**:24; solar panel installation in, **2**:25; solar tower of, **2**:47; wind energy in, **3**:59
Spas, **4**:50–51
Specialty transportation, **2**:113–14

Spindletop oil field, **1**:43
Spirit Lake Community School District, **3**:1, **3**:19–23, **3**:45, **5**:35, **5**:36*f*
Split systems, **4**:59
Spring Mills Elementary School, **5**:32
Spruill, Mary E., **3**:64–68
Spurlock Fossil Plant, **1**:121
Sridhar, K.R., **5**:121–22, **5**:121*f*
Sri Lanka, **3**:95
St. Anthony High School, **2**:60*f*
St. John Bosco Boys' Home, **4**:98–102, **4**:99*f*
Stanley, William, **1**:199, **2**:169, **3**:167, **4**:169, **5**:171
STAR. *See* Sweep Twist Adaptive Rotor
States: coal producing, **1**:114–15; hydroelectricity from, **3**:74*f;* oil-producing, **1**:55*f*
Stationary bicycles, **5**:9
Stationary fuel cell systems, **2**:119
STE. *See* Solar thermal electric
Steam reformation, **2**:91
Steam turbines, **4**:38
Steel industry, **1**:20–21
Step-up transformers, **1**:17
Stirling, Robert, **2**:41
Stirling Energy Systems, **2**:41, **2**:43*f*
Stirling engine, **2**:40–42, **2**:40*f,* **2**:56
Stokkur Geysir, **4**:38*f*
Storage tanks, **2**:77
Stored mechanical energy, **1**:5
Strassman, Fritz, **1**:134
Strategic Petroleum Reserve, **1**:60–62, **1**:60*f*
Student Energy Audit Training (SEAT), **1**:28
Students, **4**:121
Sugarcane field, **4**:86*f*
Sulfur dioxide, **1**:53, **1**:111, **1**:119, **4**:92
Sulfuric acid (H_2SO_4), **1**:110–11
Sulfur oxides, **4**:117
Sullivan, Jim, **2**:98

> 1: Oil, Natural Gas, Coal, and Nuclear
> 2: Solar Energy and Hydrogen Fuel Cells
> 3: Wind Energy, Oceanic Energy, and Hydropower
> 4: Geothermal and Biomass Energy
> 5: Energy Efficiency, Conservation, and Sustainability

Summerfield Elementary School, **5**:35–36
Sun, **1**:5, **2**:49, **2**:59, **2**:107
SunCatcher solar disk thermal system, **2**:41–42, **2**:43*f*
Suncor (Sunoco) Energy Inc, **1**:216, **2**:186, **3**:184, **4**:186, **5**:188
SunEdison, **5**:35
SunPower solar array, **2**:18*f*
Surface mining: environmental issues and, **1**:108; reclamation and, **1**:107–8; shallow coal, **1**:109
Surface Mining Control and Reclamation Act, **1**:107
Sustainable development: biofuels in, **5**:120–22; construction, **5**:82; with CSP, **5**:113; defining, **5**:100; energy efficiency and, **5**:107; geothermal energy in, **5**:109–10; goals of, **5**:100–101; green school building for, **5**:26; hydroelectric power in, **5**:108–9, **5**:108*f*; hydrogen fuel cells in, **5**:116–19; nuclear energy in, **5**:110–12, **5**:111*f*; reading materials on, **1**:162–63, **2**:132–33, **3**:130–31, **4**:132–33, **5**:134–35; renewable energy powering future of, **5**:107–22; solar energy in, **5**:112–15; system changing in, **5**:105–6; wind energy in, **5**:115–16; World Summit on, **5**:101*f*
Sustainable Energy Park, **4**:23
Sweden: biomass energy source in, **4**:88; geothermal heat pumps in, **4**:72; wave energy used in, **3**:116–17
Sweep Twist Adaptive Rotor (STAR), **3**:9
Sweet Bay, **5**:105
Swept area, **3**:8
Switchgrass, **4**:82–83, **4**:82*f*
Synthetic natural gas (Syngas), **1**:125–28; global warming and, **1**:128; hydrogen fuel cells and, **1**:127–28

Tackling Climate Change in the US, **2**:80
Taconite Ridge Wind Energy Center, **3**:34
Takasago rapid-charging station, **5**:73*f*
Tanzania, **4**:102
TAPCHAN, **3**:115, **3**:116*f*
Tapered channel wave energy, **3**:116*f*
Tarkington Elementary School, **1**:26, **5**:27*f*
Taupo Volcanic Zone, **4**:44*f*
Tax revenue, **4**:27
Technologies: Barrage, **3**:107; bio, **5**:123; biogas, **4**:102–3; clean coal, **1**:121–28; dye-sensitized, **2**:11–12, **2**:12*f*; Exide, **5**:2; geothermal energy, **4**:18; Horizon Fuel Cell, **2**:127, **5**:69; hydrogen, **2**:99–100; hydrophobic nanocoating, **5**:117*f*; nano, **4**:77, **5**:114–15, **5**:114*f*, **5**:116; Nano Solar, **2**:56; natural gas drilling, **1**:77; non-silicon-based, **2**:11; nuclear energy, **1**:151–52; Ocean Power, **3**:114; in OTEC, **3**:119–22; passive solar, **2**:60–61; photovoltaic, **2**:24–25, **2**:24–26, **2**:25, **2**:30–31; Power, **2**:83; Prism Solar, **2**:33; Quantum, **2**:122; Raser, **4**:19*f*, **4**:20; science and, **1**:218, **2**:188, **3**:186, **4**:188, **5**:190; sequestration, **1**:125; tidal, **3**:124–25; tidal

fence, **3:**107; tidal power, **3:**107–8; Vestas Wind, **3:**61; Wakonda, **2:**33. *See also* Photovoltaic technology
Tehachapi Pass, **3:**32
Telecommunications, **2:**119–20
Televisions, **5:**54
Temperature: earth's interior, **4:**3, **4:**4; earth's underground, **4:**57; hydrogen and, **2:**87; kinetic energy and, **1:**6; solar collectors and, **1:**12; solar panels influenced by, **3:**64*f;* water differences in, **1:**15
Tennessee Valley Authority Act, **3:**78, **3:**79
Terminator devices, **3:**115–16
Terrain, for microhydroelectric power plants, **3:**93
Tesla, Nicola, **1:**199, **2:**169, **3:**167, **4:**169, **5:**171
Tesla Motors, **5:**71–72
Tessera Solar, **2:**41
Texas: Austin, **5:**91; carbon footprint reduction in, **5:**11; environmentally responsible design projects in, **5:**31–32; as oil-producing state, **1:**55*f;* wind energy in, **3:**30–32, **3:**54; wind farms in, **3:**32–33
Texas Interconnection, **5:**119
Texas State Technical College, **3:**47
Thackeray, Michael, **1:**207, **2:**177, **3:**175, **4:**177, **5:**179
Thailand, **4:**39–40, **4:**87
Thames and Cosmos solar oven, **2:**73*f*
Thermal decomposition, **1:**74
Thermal energy, **1:**6, **3:**118–23, **4:**15
Thermal energy storage system, **2:**39
Thermal recovery, of oil, **1:**45
Thermochemical hydrogen, **2:**92
Thermo power plant, **4:**19*f*
Thick-film silicon cells, **2:**10
Thin-film lithium-ion battery, **5:**123, **5:**123*f*
Thin-film solar cells, **2:**9–10, **2:**9*f*
Thompson, Asa, **4:**5

Three Gorges dam project, **3:**83–85, **3:**84*f*
Three Mile Island, **1:**132, **5:**111
Tidal fence technologies, **3:**107
Tidal mill, **3:**106
Tidal power, **3:**103–4; benefits of, **3:**112; in China, **3:**110; countries using, **3:**109–11; economics of, **3:**111; energy, **1:**xvi–xvii, **2:**xvi–xvii, **3:**xvi–xvii, **4:**xvi–xvii, **5:**xvi–xvii; environmental issues in, **3:**112–13; in France, **1:**xvi–xvii, **2:**xvi–xvii, **3:**xvi–xvii, **3:**109, **3:**109*f,* **4:**xvi–xvii, **5:**xvi–xvii; functioning of, **3:**106; Golden Gate Bridge and, **3:**111, **3:**112*f;* history of, **3:**106; hydroelectric energy created by, **3:**104; in New Zealand, **3:**110–11; potential sites of, **3:**111; in Rance estuary, **3:**106; in South Korea, **3:**110; technology types in, **3:**107–8; tide differences required for, **3:**105*f;* in US, **3:**108–9
Tidal technologies, **3:**124–25
Tidal turbines, **3:**107, **3:**107*f*
Tides, **3:**104–5, **3:**105*f*
Time line, of energy, **1:**197–201, **2:**167–71, **3:**165–69, **4:**167–71, **5:**169–73
Tirevold, Jim, **3:**19–23
Titanium dioxide, **2:**13
Tiwi, **4:**41
Toledo Zoo, **4:**67
Toluene, **1:**53
Toshiba Corporation, **4:**35, **5:**118*f*
Tower, **3:**11
Toyota, **2:**86, **2:**114, **5:**69, **5:**119; FCHV of, **2:**109; RAV4, **5:**18*f*
Traeger, Tom, **2:**18–21
Traffic jams, **1:**7*f*
Transformers, step-up, **1:**17
Transmission grid: for electricity, **1:**17; electric power, **3:**25; renewable energy, **2:**28–29; for solar

1: Oil, Natural Gas, Coal, and Nuclear
2: Solar Energy and Hydrogen Fuel Cells
3: Wind Energy, Oceanic Energy, and Hydropower
4: Geothermal and Biomass Energy
5: Energy Efficiency, Conservation, and Sustainability

energy, **2**:28–29; wind energy limitations in, **3**:68–69; of wind turbines, **3**:9–11
Transportation: of coal, **1**:112–13; fuel cell applications for, **2**:107–16; hydrogen fuel cell application for, **5**:117–18; hydrogen fuel cells specialty, **2**:113–14; pipeline, **1**:77–78
Transuranic nuclear waste (TRU), **1**:146–47
Trash into Trees program, **5**:91
Trash-to-energy plants, **1**:13
Tree hugger jobs, **1**:211, **2**:181, **3**:179, **4**:181, **5**:183
Trees, capturing carbon dioxide, **5**:1, **5**:91
TRU. *See* Transuranic nuclear waste
Trucking, **1**:201, **2**:171, **3**:169, **4**:171, **5**:173
Tucson, Arizona, **5**:17
Turbines, **3**:81–82, **3**:90, **3**:92–93
Turkey, **2**:116, **4**:41
Twenhofel Middle School, **5**:32
20% Wind Energy by 2030, **3**:17, **3**:24
21st Century Green High Performing Public Schools Facilities Act, **5**:27–28
TXU Energy Solar Academy, **2**:52, **2**:54

Ulba Metallurgical Plant, **1**:141*f*
Underground mining, **1**:108–9
Underwater seabed turbines, **3**:108
Underwater turbines, **3**:110
United Arab Emirates, **5**:92–93
United Kingdom, **2**:118; FCV's in, **2**:110; geothermal power plant in, **4**:41–42; natural gas consumption of, **1**:84; wind farms in, **3**:58–59
United Nations Convention on Climate Change, **1**:32*f*
United States (US): biodiesel in, **5**:122; biofuel consumed in, **1**:61; biomass percentage used in, **4**:85; CO_2 emissions of, **1**:120*f*; coal industry of, **1**:113–15, **1**:115*f*, **1**:117; crude oil imports of, **1**:54–56; economic stimulus Bill of, **1**:33; electrical grid system improvement needed in, **5**:119; electrical grid system of, **2**:29; electricity infrastructure modernizing of, **3**:68–69; energy consumption of, **1**:19*f*; energy history of, **1**:3–4, **1**:4*t*; energy supply of, **1**:14*f*; FCVs in, **2**:107–9; geothermal companies in, **4**:20; geothermal energy in, **1**:xvii, **2**:xvii, **3**:xvii, **4**:xvii, **4**:13–20, **4**:31, **5**:xvii, **5**:110*f*; geothermal heat pump's installed capacity in, **4**:70; geothermal heat pump use of, **4**:59, **4**:70–73; geothermal resources in, **5**:110*f*; green cities in, **5**:89–92, **5**:92*t*; greenhouse gas emissions reduction target of, **5**:16; hydroelectric energy in, **3**:74–77, **3**:74*f*; hydrogen fuel cell buses in, **2**:111; hydrogen fuel cell funding cut by, **2**:86; microhydroelectric power plants potential in, **3**:96, **3**:98; natural gas consumption of, **1**:83; new oil fields in, **1**:46; Northeastern, **3**:36–37; nuclear energy in, **1**:132–33, **1**:137, **5**:112; nuclear reactors in, **1**:136; oil imports of, **1**:201, **2**:171, **3**:169, **4**:171, **5**:173; oil shale deposits in, **1**:57*f*;

petroleum consumed in, **1:**60; petroleum imported by, **1:**41; refined products importing of, **1:**59; solar energy projects of, **2:**21–23; solar energy used in, **2:**17–21; solar installations in, **2:**3–4; solar radiation across, **2:**5; tidal power in, **3:**108–9; waste vegetable oil in, **4:**114; wind energy capacity of, **3:**17–18, **5:**115; wind energy production of, **3:**29–38. *See also* specific states

United States Geological Survey (USGS), **1:**167, **2:**137, **3:**135, **4:**137, **5:**139

Uranium: energy created by, **1:**139; nuclear fuel mining of, **1:**138–40; oxide U-235, **1:**140–41; pellets, **1:**141; processing, **1:**140; U-238, **1:**144

US. *See* United States

USDA Southern Research Station, **1:**216, **2:**186, **3:**184, **4:**186, **5:**188

USGBC. *See* US Green Building Council

US Green Building Council (USGBC), **5:**26–27, **5:**36, **5:**42

USGS. *See* United States Geological Survey

Utah, **1:**109, **4:**19–20

Utility bills, **5:**36

Utility company, **3:**20–21

Utility grid: connecting to, **2:**14–15, **3:**43; solar energy and, **2:**14–15, **2:**28

Van Buren Elementary School, **5:**11

Vegetable oil, **4:**113; biodiesel as, **4:**116*f*; as fuel, **4:**114; vehicles powered by, **4:**122–25; waste, **4:**114, **4:**123

Veggie Van, **4:**127

Vehicles: biodiesel powered, **4:**110–11, **4:**117–20, **5:**75; electric, **5:**18*f*; fleet, **1:**85–86, **1:**90; fuel-cell, **5:**69–70; fuel-cell hybrid, **2:**109; gasoline use of, **1:**41; green, **5:**68–76; hydrogen, **2:**116; plug-in electric car conversion of, **5:**125–26; school, **4:**117–20; solar powered, **2:**45–46; sulfur oxides emissions of, **4:**117; using natural gas, **1:**85–86, **1:**89–92, **5:**75; vegetable oil powering, **4:**122–25. *See also* Electric vehicles; Fuel cell vehicles

Velling Mærsk-Tændpibe wind power plant, **3:**59

Verdant Power, **3:**104

Vermont, **5:**10

Verne, Jules, **2:**85

Vertical-axis turbines, **3:**13–14, **3:**13*f*, **3:**14–15

Vertical ground loops, **4:**64

Vestas Wind Technology, **3:**61

Vidaca, Jasmine, **1:**50

Vietnam, **4:**87

Villaraigosa, Antonio, **4:**34

Virginia: biodiesel school buses in, **4:**120; carbon footprint reduction in, **5:**11

Vocational information, **1:**212, **2:**182, **3:**180, **4:**182, **5:**184

VOCs. *See* Volatile organic compounds

Voith Hydro, **3:**74

Voith Siemens Hydro Power Plant, **5:**108*f*

Volatile organic compounds (VOCs), **1:**97, **1:**119

Volcanoes, **4:**7*f*, **4:**33, **4:**35, **4:**37, **4:**40, **4:**44

Volkswagen, **2:**111

Wakonda Technologies, **2:**33

Waldpolenz Solar Park, **2:**25

Walters, Bob, **1:**86–89, **1:**87*f*

1: Oil, Natural Gas, Coal, and Nuclear
2: Solar Energy and Hydrogen Fuel Cells
3: Wind Energy, Oceanic Energy, and Hydropower
4: Geothermal and Biomass Energy
5: Energy Efficiency, Conservation, and Sustainability

Washington: carbon footprint reduction in, **5:**11; Seattle, **5:**89*f;* wind farms in, **3:**36
Waste, **5:**51
Waste veggie oil (WVO), **4:**123
Water: collection, **2:**63; conservation of, **2:**65; cooling buildings with, **4:**64; temperature differences in, **1:**15
Waterfront Office Building, **4:**70
Water Furnace, **4:**75
Water heaters: conventional, **5:**67*f;* geothermal heat pump, **4:**62–63
Watermill, **1:**197, **2:**167, **3:**165, **4:**167, **5:**169
Watt, James, **1:**198, **2:**168, **3:**166, **4:**168, **5:**170
Watts Bar Unit 1, **5:**113
Wave energy, **3:**104; AquaBuoy converting, **3:**117*f;* benefits and challenges of, **3:**118; converter, **3:**115; countries using, **3:**116–18; harnessing, **3:**113–16; Norway using, **3:**117–18; from oceans, **3:**113–18; offshore generation systems for, **3:**114; onshore systems for, **3:**114–16; Portugal using, **3:**117; reading materials on, **1:**161, **2:**131, **3:**129, **4:**131, **5:**133; Sweden using, **3:**116–17; tapered channel, **3:**116*f*
Waves, **3:**113
Wessington Springs Wind Farm, **3:**36
Western Interconnection, **5:**119
Westinghouse Electric Company, **1:**145
Westlake, Mark, **2:**43–46, **2:**44*f*
Weston Solutions, **5:**40
West Virginia, **5:**32
Westwood Elementary School, **5:**28–31
Wetland water treatment systems, **1:**111
Wet-milling process, **4:**95, **4:**95*f*
Who Killed the Electric Car, **5:**71
Wibberding, Lonnie, **5:**61
Wilkinson, Martin, **3:**34
Williston Northampton School, **4:**66*f,* **4:**68
Wilmington oil field, **1:**55
Wind energy, **1:**xv–xvi, **1:**12, **2:**xv–xvi, **2:**21, **3:**xv–xvi, **4:**xv–xvi, **5:**xv–xvi; benefits and issues with, **3:**18; in California, **1:**200, **2:**170, **3:**168, **4:**170, **5:**172; challenges facing, **3:**24; compresses-air storage with, **3:**69–70; cost of, **3:**17*f,* **3:**44, **3:**68; Denmark's production of, **3:**56–58, **3:**58*f,* **5:**115; DOE report on, **3:**17, **3:**24, **3:**71; economics of, **3:**16–18; electricity generated by, **3:**38; energy storage and, **3:**69; in Europe, **1:**xv–xvi, **2:**xv–xvi, **3:**xv–xvi, **3:**57, **4:**xv–xvi, **5:**xv–xvi; farmers and, **3:**38, **3:**39; future of, **3:**24; global capacity of, **3:**56; history of, **3:**2–3; in India, **3:**59–60; KidWind project and, **3:**47–51; nanotechnology used in, **5:**116; in Northeast US, **3:**36–37; offshore power plants for, **3:**62–64; from Portsmouth Abbey monastery, **3:**37; in Portugal, **3:**59; production, **3:**29–38; reading materials on, **1:**161, **2:**131, **3:**129, **4:**131, **5:**133; Russia's potential of, **3:**62; in schools, **3:**35*t,* **3:**44, **3:**46; small residential systems using,

3:40–44; Southeast Asia sites for, **3:**61; in Spain, **3:**59; Spirit Lake Community School District using, **3:**1, **5:**35; in sustainable development, **5:**115–16; in Texas, **3:**30–32; transmission limitations of, **3:**68–69; US capacity of, **3:**17–18, **5:**115; US production of, **3:**29–38; wind speed determining, **3:**5–6
Wind Energy for Homeowners, **3:**26, **3:**72
Wind farms, **3:**15; Altamont Pass, **3:**32, **3:**33*f;* Buffalo Ridge, **3:**34; business of, **3:**39–40; in Colorado, **3:**34; compressed air storage for, **3:**31–32; electric power transmission system for, **3:**25; farmers and, **3:**39; Fenton, **3:**34; in Germany, **3:**53, **3:**57; Greensburg, **3:**34; High Winds Energy Center, **3:**32–33; Horse Hollow Wind Energy Center, **3:**31, **3:**31*f;* in Kansas, **3:**34; in Minnesota, **3:**34; in Missouri, **3:**36; Rock Port, Missouri with, **3:**29; at sea, **3:**57; in South Dakota, **3:**36; in Texas, **3:**32–33; in United Kingdom, **3:**58–59; in Washington, **3:**36; Wessington Springs, **3:**36
Wind for Schools program, **3:**44, **3:**46, **3:**46*f*
Wind generators, **3:**51
Windmills, **3:**2, **3:**26
Windows, **5:**33, **5:**51, **5:**52*f*
Wind Power in the United States, **3:**30
Wind Resource Assessment Handbook, **3:**72
Winds: basics of, **3:**3–5; direction and speed of, **3:**5; global, **3:**3*f*
Wind speed, **3:**5–6, **3:**8–9
Wind turbines, **1:**xvi, **2:**xvi, **3:**xvi, **4:**xvi, **5:**xvi; airborne, **3:**62; Bahrain World Trade Center using, **3:**55–56, **3:**55*f;* bats killed by, **3:**60; benefits and issues with, **3:**22–23; blade design of, **3:**9; in China, **1:**201, **2:**171, **3:**60, **3:**60*f,* **3:**169, **4:**171, **5:**173; companies supplying, **3:**20; components of, **3:**10; cut-in speed of, **3:**21; Darrieus, **3:**13–14; deepwater floating, **3:**63; defining, **3:**7; electricity generated by, **3:**16, **3:**21–22; energy production of, **3:**42; FloDesign, **3:**56; future uses of, **3:**72; gearbox of, **3:**9–11; generators of, **3:**9–10; for homes, **5:**63; horizontal-axis turbine, **3:**12–13, **3:**13*f;* maintenance, **3:**41; manufacturers, **3:**16; new v. old, **3:**33; Pickens plan of, **3:**27; for residential systems, **3:**40–42; residential systems maintenance of, **3:**41; revenues from, **3:**38; for schools, **3:**16–23; Skystream 3.7, **3:**46*f;* small, **3:**15; Spirit Lake Community School District with, **5:**36*f;* Texas State Technical College and, **3:**47; transmission and gearbox of, **3:**9–11; vertical-axis turbine, **3:**13–15, **3:**13*f;* world's first, **3:**2; yawing of, **3:**11–12
Wind vane, **3:**11
Wisconsin, **3:**78, **4:**17, **4:**66–67
Wood, gasification of, **4:**97
Wood alcohol, **2:**89
Wood-burning boilers, **4:**105
Wood-burning cooking stove, **4:**88
Woods, Mel, **2:**18
World Commission on Environment and Development, **5:**100
World Nuclear Association, **5:**112
World oil, **1:**190, **2:**160, **3:**158, **4:**160, **5:**162
World production, of petroleum, **1:**xi–xii, **2:**xi–xii, **3:**xi–xii, **4:**xi–xii, **5:**xi–xii

> 1: Oil, Natural Gas, Coal, and Nuclear
> 2: Solar Energy and Hydrogen Fuel Cells
> 3: Wind Energy, Oceanic Energy, and Hydropower
> 4: Geothermal and Biomass Energy
> 5: Energy Efficiency, Conservation, and Sustainability

World Resource Institute, **1:**167, **2:**137, **3:**135, **4:**137, **5:**139
World Summit on Sustainable Development, **5:**101*f*
Worldwide Fuel Cell Industry, **2:**106
Worldwide uses: of energy, **1:**19–21; of nuclear energy, **1:**136–38
Wrangell Mountains, **4:**18
WVO. *See* Waste veggie oil
Wyoming, **1:**114, **1:**115*f*, **4:**4

Xeriscape, **5:**87
Xtreme Power and Clairvoyant Energy, **1:**216, **2:**186, **3:**184, **4:**186, **5:**188

Yangtze River, **3:**83, **3:**84*f*, **3:**85
Yawing, of wind turbines, **3:**11–12
Yellowcake, **1:**140
Yellowstone aquifer, **4:**67
Yellowstone National Park, **4:**4, **4:**4*f*, **4:**9, **4:**52
Yestermorrow Design/Build School, **2:**33
Youth Awards for Energy Achievement, **3:**67–68
Yucca Mountain, **1:**148–49

Zero-carbon energy source, **1:**131–32
Zero emissions, **1:**122, **5:**73–74, **5:**94

About the Author

JOHN F. MONGILLO is presently a middle-school science teacher at Mercymount Country Day School in Cumberland, Rhode Island. He has a BS in general education, a BS in special education, and an MS in science education. John has been a coauthor and author of several Greenwood reference books, including *Teen Guides to Environmental Science, Environmental Activists, Encyclopedia of Environmental Science,* and *Nanotechnology 101.* He is also a coauthor of *Reading about Science,* a seven-book series published by Phoenix Learning Resources. He is a member of the National Science Teachers Association and the Autism Society of America. John drives a 1998 Saturn four-door sedan that was converted into a 100 percent electric plug-in vehicle by two of his students and a team of family members and technicians.